W9-ABH-573

THE UNIVERSE

THE UNIVERSE

—ORDER WITHOUT DESIGN—

CARLOS I. CALLE

59 John Glenn Drive
Amherst, New York 14228-2119

Published 2009 by Prometheus Books

Inquiries should be addressed to
Prometheus Books
59 John Glenn Drive
Amherst, New York 14228-2119
VOICE: 716-691-0133, ext. 210
FAX: 716-691-0137
WWW.PROMETHEUSBOOKS.COM

13 12 11 10 09 5 4 3 2 1

Library of Congress Cataloging-in-Publication Data

Calle, Carlos I.
The universe—order without design / Carlos I. Calle.
p. cm.
Includes bibliographical references and index.
ISBN 978-1-59102-714-0 (hardcover : alk. paper)
1. Cosmology. 2. Evolution (Biology) 3. Life—Origin. 4. Chaotic behavior in systems. I. Title.

QB981.C345 2009
523.1—dc22

2009005252

Printed in the United States on acid-free paper

To Benjamin,
who may live to see the answers to many of our questions.

CONTENTS

Preface

During a scientific conference held shortly after the introduction of the inflationary theory describing the first moments of the universe, I asked Howard Georgi of Harvard University, whom I had just met, for a quick explanation of the negative pressure idea that was central to the new theory. Georgi promptly told me that he didn't understand that concept either and suggested that I should ask Alan Guth, the originator of the theory, who was also at this conference. I never had a chance to ask Guth directly and it was only after careful study of the papers and books that were eventually published that I was able to get a feel for the idea.

My goal with this book is to explain not just this idea of a universe that starts out with negative pressure but also all the revolutionary concepts behind the new scientific theories that are taking us beyond the moment of the big bang. These exciting theories and models are beginning to describe for us a universe that was never born and will never die, a universe that is fully explained by science.

In writing this book, I had the assistance of many people. My thanks go first to my wife, Luz Marina, for her understanding and unselfish support during the many hours that I spent at the keyboard. I would also like to thank my son Daniel for his enthusiasm about the central idea of this book. I wish to thank Professor Matt Young, of the Colorado School of Mines, who carefully and thoroughly read the entire book and made helpful suggestions and corrections. I would also like to thank Kristine Hunt, who provided meticulous editing and who made sure that the manuscript was consistent. I am grateful to my agent, Susan Ann Protter, for her kind support throughout the entire project. Finally, I would like to thank my editor, Linda Greenspan Regan, for her efficient editing.

Chapter 1

DESIGNER UNIVERSE

PHOTONS FROM THE SUN

The light reaching your eyes after it is reflected by the light areas of the page that you're reading now originated a million years ago, deep in the interior of the sun, where the temperature reached 10 million degrees. Some 600,000 km (375,000 miles) beneath the gaseous, hot, bright surface that we see from our planet, the cores of two hydrogen atoms coalesced into the nucleus of a heavy form of hydrogen called deuterium. A few moments later, the newly formed deuterium nucleus collided with a third hydrogen nucleus to form helium. The second collision produced something else: a photon, a tiny bundle of light that immediately started on a long and fortuitous journey toward the sun's surface. Life wasn't easy for that photon. Countless times, atoms and electrons in the sun absorbed the photon only to regurgitate it almost immediately, not without taking away some of its energy. More times than you'd care to count, the photon got ejected in the wrong direction, losing most of the ground that it had gained in previous attempts.

A million years after it started on its journey, the photon finally reaches the surface of the sun and sets out toward the Earth. Eight minutes later it is absorbed and reemitted by the molecules that make up the paper of this book. An instant later, the photon comes out and enters the cornea of your eye. As it did in the interior of the sun and while traveling through the Earth's atmosphere, the photon once again is absorbed and reemitted by the molecules in the cornea and the lens.

The ciliary muscles that surround the lens relax, decreasing the radius of curvature, sending the photon through the vitreous, jelly-like body toward the paper-thin retina. There, the retinal, a compound derived from vitamin A, absorbs the photon, ending its million-year journey.

The energy of the photon absorbed in the back of your eye alters the geometry of the retinal, and this change causes a series of molecular transformations that triggers an electric signal that is carried by the optic nerve toward your brain. When the electric signals from a few other photons that originated a million years ago in the depths of the sun follow the first one, your brain integrates these signals and begins to form the image of the words on this page.

The actual processes both in the eye and in the core of the sun are much more complicated than described here. At first glance—a physicist's glance, but a glance nonetheless—the thermonuclear reactions that take place in the sun shouldn't happen. The hydrogen nuclei have positive electric charges (they are actually single protons) that repel each other with a force that becomes extremely large when they are brought close to each other. The two colliding hydrogen nuclei don't have enough energy to overcome this electrical repulsion, even when moving at extremely large speeds near the center of the sun. How do they do it? They actually tunnel through the electrical barrier. It is as if you were repeatedly bouncing a tennis ball against a wall with your racket, hitting the wall higher and higher, when suddenly, a few feet before the ball reaches the top, the ball goes through the solid wall and appears at the other side. According to quantum mechanics, that's exactly what happens to the colliding protons in the sun so that they can create the photons that eventually enable you to read this page.

Your eye is able to detect about five consecutive photons in a relatively dark room. In the brighter room you're sitting in now, you are actually receiving millions of photons. But they have to be the right photons; they have to have the right energies or you wouldn't be able to see them. The photons that are created in the sun have way too much energy for your eye to detect them. If fact, they have so much energy that they would damage your eye. They are actually gamma rays, a type of radioactivity. It is through their very frequent collisions that they lose the exact amount of energy so that your eye can detect them. And it takes a million years of continuous collisions to achieve that.

WAS THE SUN DESIGNED FOR LIFE?

The sun sends out photons with energies other than visible light. However, most of them are in the visible range of energies. What's more, the sun's photon output peaks in the yellow-green part of the spectrum, which happens to be where our eyes are most sensitive (Fig. 1–1).

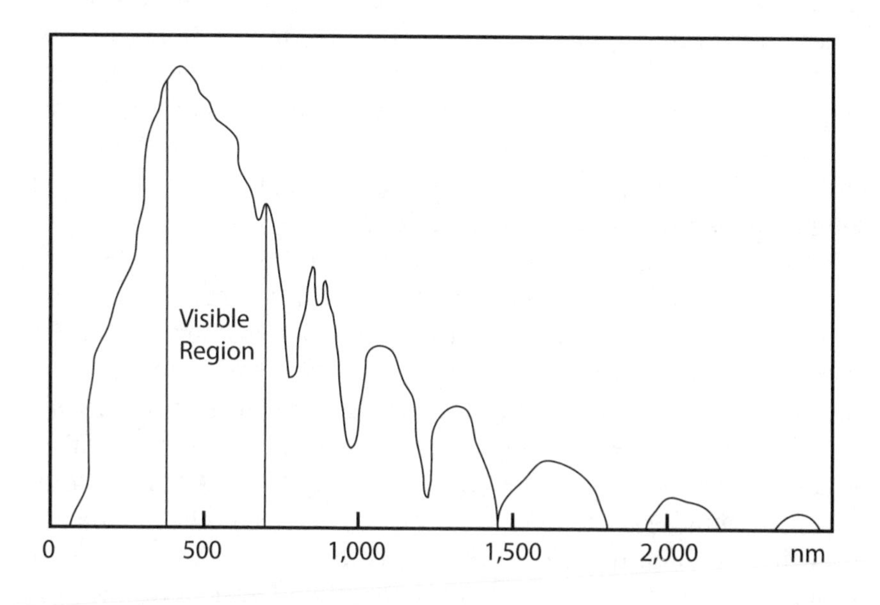

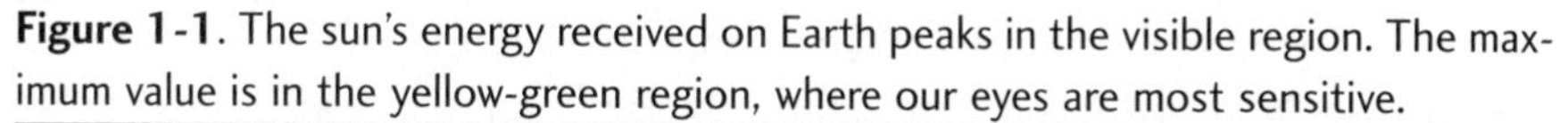

Figure 1-1. The sun's energy received on Earth peaks in the visible region. The maximum value is in the yellow-green region, where our eyes are most sensitive.

There seems to be a remarkable equilibrium between the complicated mechanism of light and energy production in the sun and human vision. But this is just one of the many equilibriums that exist between the sun and the Earth. The most important one is that the total energy output from the sun allows for the existence of abundant liquid water on the Earth, an essential ingredient for life.

Is this equilibrium a coincidence or was it designed specifically for our existence?

In the fourth century, Augustine asked a similar question. To him, the order, structure, and beauty of the world required a supreme designer. Thomas Aquinas in the thirteenth century came to the same conclusion. Perhaps the most extensive description of this "Argument for Design" was put forward in the eighteenth century by the English theologian William Paley in his popular book *Natural Theology or Evidences for the Existence and Attributes of the Deity Collected from the Appearance of Nature.*[1] Paley imagined finding first a stone and then a watch while walking in a forest. If someone asked about the origin of the stone, he could easily say that it had been there forever. The watch, however, required that someone assembled it together and carefully adjusted it so that it would move with the precision needed to accurately show time. This complex piece of machinery, Paley thought, "must have had a maker, . . . an artificer or artificers who formed it for the purpose which we find it actually to answer, who comprehended its construction and designed its use."[2]

The mechanism for the generation of visible light in the sun is more delicate and precise than any watch. But the sun hasn't always been this way. The sun has evolved and continues to evolve. The sun was born 5 billion years ago in a vast cloud of hydrogen gas when interparticle collisions generated a region of higher concentration of gas. This region separated into a large sphere, or *protostar*, which soon began to shrink because of its own gravity, causing the temperature to rise. Two hundred and fifty thousand years later, the temperature reached 10 million degrees, high enough for nuclear fusion to begin. At that moment, the sun was born. Fifty million years after the protostar was formed, the sun reached its maturity stage.

During the sun's adult life, which lasts 10 billion years, most of its energy comes from the conversion of hydrogen into helium. During that time, the temperature increases steadily. Our sun is now middle-age, having lived 5 billion years, with another 5 billion years left before its nuclear fuel runs out in the core. When that happens, the core won't be able to hold its own weight and will start to contract. The contraction will heat up the core again and the sun will start radiating energy out. The outer layers will expand, turning the sun into a *red giant*. The Earth by then would long be deserted—which is a good thing, since the red giant sun will engulf it along with Venus and Mercury.

The red giant phase of the sun will last for a few hundred million years, after which gravity will start to compress the core of the sun until it lights up once again as a *white dwarf*. This third phase in the life of the sun will last for a few million years. Its end will now be near, as the sun becomes a *black dwarf* with no fuel to burn—a dead star.

The sun is the perfect star for life only during most of its adult years, when the temperature remains close to 6,000 kelvin (or 6,000 K).[3] Life as we know it, and even life as we may be able to imagine it based on the properties of its possible constituents, is unlikely to be possible during the red giant or white dwarf phases. And it will certainly not be possible during the black dwarf phase.

This complex watch, the delicate and precise mechanism of energy and light production in the sun, is crucial for our existence. Now, I'd like to tell you about another watch.

Star Stuff

The selenium, copper, and zinc in your body all come from massive stars that exploded over 5 billion years ago. As you might imagine, star explosions aren't an everyday event. Every forty years or so one such star explosion—a *supernova*—occurs in our galaxy, but most are hidden by interstellar dust. In fact, the last supernova seen in our galaxy was discovered by Kepler in 1604. But the universe has billions of galaxies, and in recent years, one supernova explosion took place in a nearby galaxy right before our eyes and telescopes.

On February 24, 1987, astronomer Ian Shelton of the University of Toronto was taking photographs of the Large Magellanic Cloud with the telescope of Las Campanas Observatory in Chile. The Large Magellanic Cloud is a small satellite galaxy of our own Milky Way galaxy. The night that followed, he continued with his project of photographing the same region of space. After examining the new photographs and comparing them to the ones he'd taken twenty-five hours before, he noticed a new bright star on the new plates that wasn't on the earlier plates. The new star turned out to be the first supernova explosion observed since Kepler's (see Plate I).

In New Zealand, amateur astronomer Albert Jones actually saw the explosion directly. One evening, just before clouds rolled in, Jones noted a "bright blue object" in the Large Magellanic Cloud. He was able to confirm the event later that evening when the sky cleared.

Jones saw it in 1987, but the explosion actually took place 167,000 years ago, since the Large Magellanic Cloud is 167,000 light-years away. When you look at a red sun during a beautiful sunset, you're looking at the sun as it was 8 minutes ago, since that's how long light takes to get here from the sun. When you look at your friend across the table, you're looking at the way she was 3 nanoseconds ago because that's how long light takes to bounce off her and reach your eyes. Of course, your friend doesn't change that much in 3 nanoseconds and when you look at her, that's the way she is now. The sun doesn't change that much in 8 minutes, either, so the sun in the sunset is still there and hasn't exploded. But Sanduleak, the large star that exploded in the Large Magellanic Cloud, 167,000 light-years away, does. And did. When Kepler saw his supernova in 1604, Sanduleak had already exploded, but he didn't know it. He couldn't have known because the light from that explosion hadn't reached us yet. Once it reached us, it shone as bright as the stars in the Big Dipper, releasing at once the same amount of energy as one hundred suns would generate in their entire lifetimes.

The sun can't explode into a supernova. It's not big enough for that. You need stars larger than about eight suns to have supernovas. The best ones are as large as twenty-five suns. As a star burns its fuel, it generates heavier elements in its core. A massive star can generate elements from helium to iron so that when all of its fuel is used up, it ends up with a core the size of the Earth and made solely of iron. When energy production halts, there is no outward radiation to counteract the force of gravity so the star collapses quickly, taking one second to squeeze the core into a small sphere 100 km (60 miles) across!

The collapse increases the temperature of the core, providing enough energy for the production of elements heavier than iron. The core collapse eventually stops and the material rebounds outward, crashing into the plunging outer layers and forming a shock wave. The explosion takes a few tens of milliseconds, sending most of the matter that made up the star into space, including the selenium, copper, zinc, and all the newly made heavy elements that make up the Earth and part of our bodies.

We've found a more complex and delicate watch. The heavy elements that our bodies need are made in the last moments in the life of a massive star, during the complicated series of events that culminate in its destruction. And it is this destruction that makes possible our existence.

Other Watches

There are more watches to be found in nature, some of which we'll be exploring. There are complex mechanisms in place that make the existence of life on Earth possible. There seems to be a fragile equilibrium between a complex process in the distant past and our own existence. If the mechanism had been slightly different, we wouldn't be here.

It can be argued that our first watch may not be that unique. Stars come in many varieties and their core temperatures can be very different (see Plate II). Although all stars use the same mechanism of conversion of hydrogen into helium that generates the photons that the sun emits, a star with a much higher temperature will release many more of these photons, producing more energy than we would need and would make the Earth too hot for life to exist. Other stars may produce too few photons and may be too cold for life. But there are enough sunlike stars in our galaxy alone where a rocky planet similar to our Earth and orbiting at about the same distance could have the right conditions for life to appear, evolve, and survive.

However, there is one watch that can't be explained with this argument. Recent discoveries are showing that a quantity that Einstein introduced into his equations of general relativity—which he called the *cosmological constant*—has a value so perfectly fine-tuned for the emergence of life in the universe that the tiniest deviation from it would destroy the delicate balance that allows life to exist. This really is the biggest watch of all.

The reader might have noticed that my watches have all been physics watches, that I didn't bring up any biological watches at all, even though those were what Paley had in mind when he wrote his popular book. One reason for this is that I am a physicist, not a biologist.

But the main reason is that, other than the origin of life itself, all of biology has been successfully explained. Biology has a magnificent theory: Darwin's theory of evolution. And all of biology can be explained with it. Biology has a designer, a watchmaker, but it is a blind watchmaker, a mindless watchmaker without a purpose. Biology's watchmaker is natural selection. Richard Dawkins's wonderful book, *The Blind Watchmaker* explains it clearly and authoritatively.[4]

Although biology deals with what may be the most complex systems in the universe, physics concerns itself with the ultimate questions of existence: How is the universe made? How does it work? Did it have a beginning and if so how did it start? Equally important: Was the universe designed for life?

Biology can be explained through natural selection. The universe can be explained with the laws of physics, its watchmaker. But unlike natural selection, the theory that would explain the universe and how the laws of physics came about is still in development.

MODELS OF THE UNIVERSE

The path toward a theory of the universe started almost a hundred years ago with Einstein. After completing his masterpiece, the general theory of relativity—a theory of gravity that replaced Newton's own masterpiece—Einstein decided to apply it to the universe as a whole. Einstein used his field equation of general relativity to build the first model of the universe. The universe that he eventually came up with was static and, in that aspect, it fit well with the observational evidence of the time. However, Einstein's original model didn't really predict a static universe. It predicted a dynamic universe that was changing in time, either expanding or contracting. At the time, scientists thought that the universe was more or less constant in time. To them, the universe could have existed forever, although there were problems understanding how stars could continuously emit energy for an infinite time. Because if they did, the entire universe would be as hot as the interior of the stars. Still, Einstein decided to make his model fit the observations of a static universe and to that effect added a "cosmological term" to his equations. This term added a repulsive force that exactly counterbalanced gravity and forced the model to represent a static universe.

The Russian scientist Alexander Friedmann decided that Einstein's term didn't belong in the model and removed it. He wanted to see where Einstein's original model would take him. It took him to the same result that Einstein had rejected, to a universe that was either collapsing or expanding. Einstein originally opposed Friedmann's calculations but, after examining them more carefully, reluctantly admitted that they were "both correct and clarifying."[5]

Independently from Friedmann, the French scientist Georges Lemaître also proposed a model of the universe based on Einstein's equations without the cosmological term. But Lemaître went beyond Friedmann and actually predicted that Einstein's general relativity showed that the universe started as a small region that exploded and evolved into the universe we see today. In a 1927 paper he predicted that the galaxies should be moving away from each other and that their speeds should be proportional to their distances.

In 1929, a revolutionary discovery dramatically corroborated Friedmann's and Lemaître's predictions. That year, Edwin Hubble found that all galaxies in the universe were rushing away from each other and that the farther away they were, the faster they moved. Hubble discovered that Friedmann and Lemaître had been right all along: the universe was expanding. Sadly, Friedmann wasn't alive to see it. He had died of typhoid fever in 1925.[6] Lemaître was more fortunate. He saw his model developed into the more sophisticated big bang model of the origin of the universe.

Hubble's discovery showed us for the first time in history that the universe must have had a beginning. Because of this, his discovery ranks as one of the greatest of the twentieth century.

Newer Models of the Universe

Advances in our understanding of the structure of matter allowed scientists to build more precise theories of the origin and evolution of the universe. The big bang theory, developed by George Gamow and others, explained the expansion of the universe and the origin of the elements. In 1964, an accidental discovery from a microwave antenna designed to listen in on radio sources outside our galaxy proved one of the main predictions of the big bang theory. When the two scientists (Arno Penzias and Robert Wilson of Bell Labs) started their experiments, they couldn't get rid of a relatively large "noise" being detected by their antenna and that corresponded to the radiation emitted by a gas cooled to a temperature of 3 K. Initially, they attributed the noise to a "white, electrically insulating residue" left by a pair of pigeons that had taken up residence in the horn of the antenna. But the "noise" remained even after the pigeon droppings had been cleaned away.

Although Penzias and Wilson didn't know it at the time, the big bang theory had predicted that as the universe expanded, it would cool off, and that by now, its temperature should be about 3 K. Which was what Penzias and Wilson's microwave antenna was detecting. The two scientists had actually discovered the echo of the big bang, the temperature of the background radiation left over by the explosion that created the universe.

After the euphoria of this success, the theory ran into difficulties when it couldn't explain several observational discoveries that showed that, on a large scale, the universe looked the same in any direction. According to these observations, the big bang must have sent equal amounts of matter rushing away in all directions with extraordinary precision, an unlikely outcome of an explosion.

In 1981, Alan Guth at MIT proposed an advanced big bang model of the universe in which the universe undertook a very short period of an extremely rapid expansion. His inflationary model, as he has called it, resolved the uniformity problem that the big bang theory had encountered, and today it's our best theory of the evolution of the early universe.

THE PROBLEM OF ORIGINS

Discovering how the universe evolved from its early stages to the formation of galaxies and stars is a great advance, but it isn't enough.

We'd also like to know how the universe is put together, what its building blocks are, what makes it work. We're looking for a theory that explains the universe, a theory that would reveal to us the physics watchmaker. This is perhaps the area of physics where we've seen the greatest advances.

This search for the building blocks of matter and for the laws of nature brought us closer to the early history of the universe. In their search for simplicity, physicists have found that the laws of nature become simpler the farther back they look in the history of the universe. During the early moments of the universe, the temperature reached very high values. At those temperatures, the different laws that govern the behavior of matter in our universe began to merge into one another in a process of unification of the forces of nature.

Today, all of the physical processes that take place in the universe are the result of four individual forces of nature: the gravitational force; the electromagnetic force; and two nuclear forces, the weak and the strong. When the higher temperatures of earlier times are reproduced in large particle accelerators, physicists see that the electromagnetic and the weak nuclear force unite into one single electroweak force. Reproducing the temperatures of still earlier times, they are beginning to find evidence that this new electroweak force unifies with the strong nuclear force, although they don't yet know the exact details of how this next unification takes place. Our theories and experiments can reproduce the conditions of the early universe up to this point, a trillionth of a second after the big bang, but taking the last step, reaching time zero, has proven elusive.

Physics is taking us tantalizingly close to the very moment of creation. When you get that close to answering the big question of the origin of the universe, it's impossible not to ask about what happened before that moment. If the universe began in a big explosion, a big bang, what came before?

Many thinkers throughout history have tried to answer this question. Augustine's answer was that the world was created with time and not in time.[7] Time was created with the universe along with space. There was no before. As Stephen Hawking of Cambridge University says, asking about a time before the origin of the universe is like asking what lies north of the North Pole. The question has no meaning.

Since the early universe was very small, the quantum mechanical uncertainties inherent to the atomic world also applied to it. The moment of creation is then blurred in this quantum uncertainty. In an attempt to understand this epoch, Hawking and his collaborator James Hartle of the University of California at Santa Barbara treat space and time on an equal footing by applying an elegant mathematical tool used in quantum mechanics to the conditions of the early universe. In their approach, the quantum-sized universe contained only space with no edge, no boundaries. In this four-dimensional space, the universe had no beginning.

New cosmological models based on string theory and M-theory, on the other hand, tell us that the universe may have existed for an eternity, with no beginning and no end. In this new view, our universe may be part of a vast landscape of other universes that are continuously being born. In each universe, time and space begin anew.

WHAT ABOUT GOD?

If, as Augustine said two centuries ago, time and space started with our universe and there was no before, how did it all begin? Did the early universe contain only space and no time, as Hawking and Hartle propose? If this was the case, how did space appear and how did it start its partial transformation into time to create the spacetime continuum that forms our universe? Was the tiny region of space that was our universe before time got going simply sitting there for an eternity, without any changes or motion? Or, as Hawking says, is time only an illusion and does the universe—even today—contain only space?[8] We are in danger of falling into quicksand here, since we can't conceive of space without time nor can we form any image in our minds of something existing outside time. But these ideas can be expressed in the language of mathematics; we can write equations describing space that don't involve time.

In the original big bang theory, the universe has been expanding since the original explosion that created it. Imagine that the current universe is enclosed in a very large sphere, as in Fig. 1–2. If you were to cut this sphere in half, the edge of each hemisphere would be a large circle with the same diameter as that of the universe. Now, if you could run the "movie" of the expansion of the universe backward, you'd see that as the universe contracts, this circle gets smaller and smaller. If you stack these circles on top of each other, you'll have the cone shown on the right of Fig. 1–2. The sharp point at the bottom represents the origin of the universe.

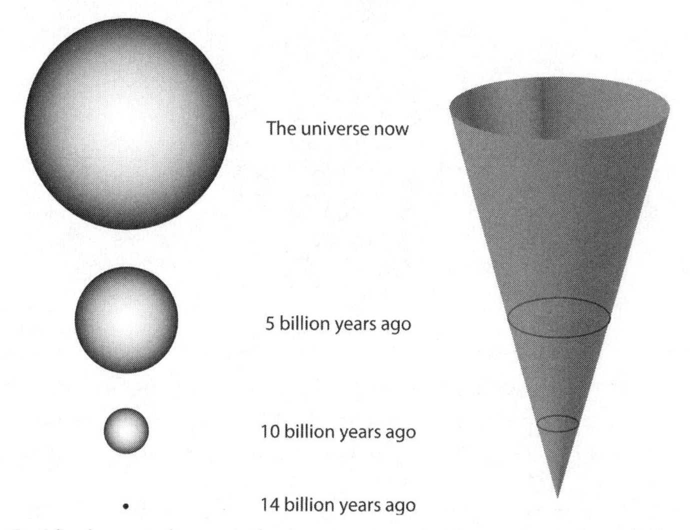

Figure 1-2. The large sphere at the top represents the current size of the universe. Running a "movie" of the expansion of the universe backward will show the universe with a smaller and smaller size until it reaches a tiny point. Stacking the circular cross sections of these spheres gives us the cone on the right. The tip of the cone represents the origin of the universe.

If we were to magnify what appears to be the sharp point at the start of the big bang, we would discover that it's not really sharp; it's a blunt, rounded tip (Fig. 1–3). The quantum mechanical uncertainty has removed the fine detail of the tip. In the model of Hawking and Hartle, the origin of the universe is no longer a well-defined event. There is no exact point of creation, no boundary. Time, treated as a space coordinate, is still finite but unbounded, like the surface of the Earth. When you travel around the Earth, you never reach an end and you never fear falling off the edge. But even though the Earth's surface is unbounded, it's not infinite and you can measure it (it's 500 trillion square meters, or 200 million square miles). That's also the case with the early universe. The rounded surface of the tip has no sharp boundary, no edge. There is no definite origin of the universe. In Hawking's words, the universe is "completely self-contained and not affected by anything outside itself. It would neither be created nor destroyed. It would simply BE."[9]

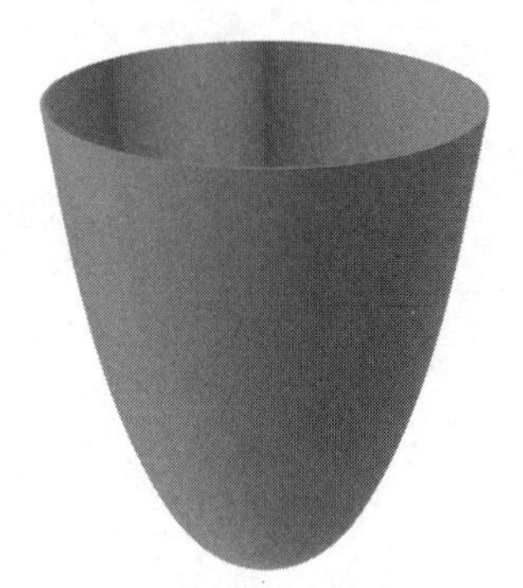

Figure 1-3. The early universe is a quantum mechanical object. The quantum uncertainty tells us that if we look closely at the tip of the cone of Fig. 1-2, we discover that there is no sharp point, but rather a blunt end. The origin of the universe is fuzzy, no longer well defined. There is no exact point representing the creation of the universe.

But the universe is ruled by the laws of physics. The question now becomes: Who created the laws of physics? Can they just simply be?

According to recent work by Hawking, working this time with Thomas Hertog of the European Center for Nuclear Research (CERN),[10] the laws of physics are chosen by living observers, human or not. In their model, the universe doesn't have a unique history. Instead, it has lived through all possible histories that started out when the universe was small enough to be a quantum mechanical system. The idea of considering all possible histories of a quantum mechanical object was first introduced by the great American physicist Richard Feynman. This idea is now a fundamental part of physics and physicists everywhere use it in performing calculations that are corroborated by experiments. To paraphrase Hawking, if their view is correct, there must be a history of the universe in which the United States has beaten Brazil in every single soccer world cup, although that's still an unlikely history.

In this new view, the universe and its laws are selected by the observers themselves. If the universe is self-contained, with no beginning and no end, as Hawking speculates, there is nothing left for a creator to do.[11]

If, instead, the new models that are based on string theory and M-theory are correct, our universe is either one of a vast number of universes that have existed for an eternity or a link in an infinite chain of universes. Although our own universe had an origin in the big bang, the fundamental laws of physics existed before.

String theory and its extension, M-theory, are theories of the structure of matter that attempt to unify all the forces of physics and that are formulated in ten dimensions of space and one of time. M-theory contains the descriptions of an estimated 10^{500} possible universes, a number so large that we don't have any way of grasping it other than mathematically. The "eternal inflation" model of the universe proposes that each one of these possible universes actually exists and gives rise to an ever-increasing number of pocket universes within itself, a process that continues forever. These pocket universes begin with a big bang and a brief period of rapid inflation during which they increase in size from a tiny volume much smaller than an atom to a volume larger than our visible universe. While many of those universes may contain stars and galaxies and perhaps life, many more may be ruled by strange laws of physics that produce chaotic worlds.

The fundamental ingredients of M-theory are multidimensional membranes, or *branes*. According to what is called the eternally oscillating model, the entire universe exists in a spacetime of two multidimensional branes that enclose an eleven-dimensional region. One brane contains the stars and galaxies of our universe and the other, located a fraction of a millimeter away, houses a parallel universe with its own separate worlds. The two universes repeatedly collide and recoil in cycles that last a trillion years. Each collision starts the cycle of a new universe that begins with its own big bang and evolves stars and galaxies. If this model is correct, we are living in one of the cycles. A trillion years from now, our universe will collide with its parallel universe on the other brane, ending in a big crunch that will produce a new universe, with its own set of laws of physics, and its own stars and galaxies. The series of cycles has been in existence for eternity and will continue forever.

If either the eternal inflation or the eternally oscillating model is correct, the creator doesn't have a job to do either.

However, the delicately fine-tuned value of what we referred to as our biggest watch—the cosmological constant—is so extreme that scientists are having a very difficult time explaining it fully with any model. Does the creator have a job now? Is a supernatural creator needed to choose and adjust the cosmological constant for our universe so that it can harbor life? Let's explore this tantalizing question in the pages that follow.

Chapter 2

Building a Theory of the Universe

Explaining the Universe

To the ancient Babylonians who lived in Mesopotamia some five thousand years ago, the origin of the universe had been fully explained: In the beginning there was only water in the form of Apsu, the fresh water, and of T'iamat, the chaotic salty water of the sea. From their union came the gods of heaven and Earth, who were trapped in the body of T'iamat. The son-god Marduk slew T'iamat, cutting her body in two to separate heaven and Earth, thus removing the original chaos and creating order in the world. The blood of the slain fertilized the Earth and gave birth to the human race.

The Egyptians also knew how the universe began: Before the creation there was Nun, the primordial oceanic abyss. Nun contained Atum, a formless, yet-to-be-completed spirit that had in it all of what was to exist. This spirit took the form of the god-sun Ra, who created the universe, the gods and goddesses who ruled it, and all the living creatures.

The early Greeks also knew how the universe began. In the beginning, there was only Chaos, the infinite abyss. Then there was Gaia, the Earth; Eros, the spirit of life; and Tartarus, the lower world. Chaos created Erebus and Night, and out of their union came Aether, the sky, and Day. Gaia then created Uranus, the starry sky, as well as the sea.

Gaia and Uranus then gave birth to the first gods, the Titans, who were the first rulers of the world, and to the Cyclops and the Giants.

Sometime between 1500 BCE and 500 BCE, the Israelites came into possession of a detailed description of the way God created the universe. According to the Genesis account, adopted later by the Christians in the first century, in the beginning God created the heaven and the Earth, but the Earth was void and without form. God then created light, divided it from the darkness, and there were day and night. God separated the waters and created the firmament, which he called Heaven. He created the sun to provide light during the day and the moon to shine a lesser light at night; he also made the stars, and he placed all these lights in the firmament of the heaven to give light upon the Earth. After all this was in place, God created the whales and all the creatures that populate the rivers and the seas, and the cattle, the beasts, and the animals to populate the Earth he had just created. God then made man in his image to rule over the fish of the sea, the fowl of the air, the cattle and the beasts of the land, and over all the creatures that live upon the Earth. And having created all of this in six days, God rested on the seventh day from all the work that he had done.

Similar mythological tales that attempted to explain the universe can be found in most prescientific societies. The Minoan, Chinese, Norse, Celtic, Indian, and Mayan cultures all weaved myths into images to explain the universe that they observed. Common to most of those tales is the creation of the world from a void or a chaotic abyss into a home for mankind, which is at the center of creation. Some of the tales are elaborate and imaginative, with gods and goddesses going into battle to destroy the evil spirits of chaos so that they can create the Earth and the sun, the wind, and the rain. In other tales, sleeping gods must be awakened to infuse life into the universe. Still other tales, such as the Genesis account, detail in an anthropomorphic way the creation of the Earth, the sky, the sun, the moon, the stars, the living creatures, and finally, the human race. The simple narrative is an obvious description of the world observed by the people in their time, with the sun as a ball of fire in the sky providing light and heat to the Earth, and the stars as small lights affixed to the firmament, a dome surrounding their entire universe.

A Rational View

The path toward a rational view of the universe was long and tortuous. This path seems to have started with the agrarian civilization of the Babylonians, who required knowledge of the periodicity of the seasons to determine the best harvesting times. This need prompted them to undertake a systematic study of the heavens and, in doing so, they invented a rudimentary geometry. The Egyptians also devised geometrical concepts to be able to survey land areas and to construct their temples and pyramids. The discoveries of these early civilizations reached the Greeks, notably through Thales of Miletus (634–546 BCE), who visited Egypt and probably Babylonia, and who was very likely educated in Eastern astronomy and geometry. Thales made a fundamental advance in mathematics with his invention of the deductive method by which mathematical statements are proven through a regular series of arguments.

The ideas and concepts that Thales put forth laid the foundation for the advances in the Greeks' understanding of the world. Anaximander (610–546 BCE) discovered that the heavens revolved around the North Star and proposed that the sky was a sphere around the Earth. Noticing that the position of the stars changed as he traveled, he proposed that the Earth's surface was curved, not flat, as had been believed. Observations of the Earth's shadow cast on the surface of the moon during lunar eclipses supported this view. In the third century BCE, Eratosthenes was able to measure the circumference of the Earth by measuring the angle between the sun's rays and a stick that he stuck vertically into the ground in his native city of Alexandria (Fig. 2–1). Knowing that in the city of Syene the sun shone on the bottom of a well at noon on the first day of the summer, he used the angle he had measured in Alexandria and the distance between the two cities to obtain the value of the circumference of the Earth with remarkable accuracy.

Eratosthenes didn't stop there. With his value for the size of the Earth, he could now calculate the distance to the moon by comparing the circumference of the Earth's shadow on the moon during a lunar eclipse with the actual value of the Earth's circumference. He then compared the value for the size of the moon just obtained with the distance at which his thumb exactly blocked the disc of the moon.

From this comparison, he constructed two similar triangles that gave him the distance to the moon.

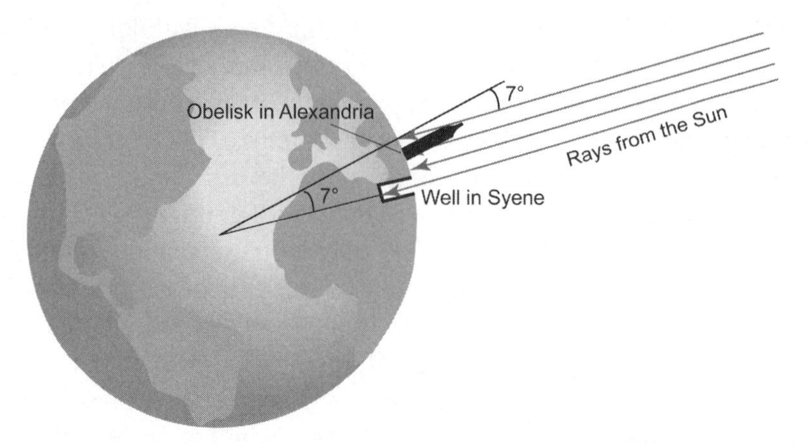

Figure 2-1. Eratosthenes measured the circumference of the Earth by measuring the angle that a vertical pole in the city of Alexandria cast at noon on the first day of summer. On that same day in Syene, located south of Alexandria, the sun's rays came in vertically and didn't cast a shadow at noon. Using simple geometry, he used the angle measured by the pole and the shadow and the known distance between Syene and Alexandria to calculate the circumference of the Earth.

Using a geometrical method proposed by Anaxagoras of Clazomenae and Aristarchus of Samos, Eratosthenes also calculated the distance to the sun as well as the sun's size. The results were only approximate due to the difficulty in determining the exact position of the sun and the Earth that were required for his calculation.

These remarkable calculations and detailed observations of the heavenly bodies gave the Greek thinkers an idea of the scale of the known universe. In the third century BCE, Aristarchus said that these observations could be explained more simply if the Earth and all the planets revolved around the sun and that the stars, motionless in the sky, were very far away. He also said that the Earth rotated on its axis as it revolved around the sun and that this axis was tilted with respect to the plane of the orbit.

Sadly, this extraordinarily modern view of the world didn't last. Aristarchus was ahead of his time and the necessary tools to validate his model would not exist for two thousand years. How could the

Earth revolve around the sun if you couldn't feel any motion? Shouldn't the stars shift in the sky as the Earth travels around the sun? Although Aristarchus had stated that the stars were far away, he didn't realize just how far they actually were. The apparent shift in the motion of the stars as the Earth revolves around the sun is extremely small and requires a telescope to be detected. Not even after the telescope was invented early in the seventeenth century was it possible to measure this apparent shift, or *parallax*, as the early telescopes didn't have enough magnification to detect this small effect. When the German scientist Friedrich Bessel finally measured the parallax of a star called 61 Cygni, the angle subtended by its apparent motion turned out to be only about 0.3 seconds of arc. Observing this small motion is equivalent to detecting a dime at a distance of ten kilometers (six miles)!

An Irrational View

With no possibility of detecting the motion of the Earth as it moved around the sun, Aristarchus's model wasn't accepted. Instead, the traditional view with the Earth at the center of the universe that Plato and Aristotle had promulgated was embraced with new fervor. In the fifth century BCE Plato had decided that since the heavens were perfect, the heavenly bodies must move in perfect paths, and the perfect path was the circle. Aristotle accepted his teacher's idea and said that the heavenly bodies must move in steady, permanent, and perfect circles.

Although the sun, moon, and stars appeared to move in the steady circular motion that Aristotle claimed, the planets didn't. They moved in circles but not in the steady, even way that Aristotle taught. To explain the seemingly complicated motion that results, as we know today, from the combination of the planets' own motions and the Earth's motion around the sun, Claudius Ptolemy, in the second century CE, developed a complex model of the universe. He had the planets moving in small circles that in turn would revolve around the Earth in their own circular orbits. When more accurate observations of the motion of the planets were obtained, Ptolemy simply added more circles that moved around other circles, with each planet on the last one. Eventually, Ptolemy's system had forty circles with which he was able to reproduce the observations.

The more rational view that Eratosthenes and Aristarchus had developed was replaced by irrational views based on the mythological belief that the heavens must be perfect, eternal, and immutable. Since the sun, moon, planets, and stars were observed to move in circles around the Earth, the circle must be a perfect geometrical figure. If the planets didn't seem to obey the rule and moved in apparently erratic ways, circular motion was imposed on them because, being heavenly bodies, they must be perfect and move in perfect ways. When new observations didn't fit this perfect circular path, more circles were forced on the model, regardless of how cumbersome it became.

Historical events influenced the extremely long survival of the Earth-centered view of the universe. Ptolemy developed his model during the period 127 to 150 CE, at the time when Christianity was emerging as one of the world's great religions in the cradle of the Roman Empire, itself enjoying an unprecedented period of peace, stability, and prosperity. When Christianity was firmly established, the Roman Empire entered its period of decline, eventually falling to the barbarian invasions from northern and eastern Europe. During these periods of turmoil, the Christian church was the only institution capable of providing stability and unity, replacing the authority vacated by the Roman governors who had been removed. Thus the role of the church in human affairs, religious or secular, grew to unprecedented levels. During the High Middle Ages (1000–1300 CE), the period of political expansion in Europe, the church, with its power now concentrated on the pope in Rome, established its authority independently from the individual monarchies. The papacy became an independent political power permeating most of Europe.

The church found that the teachings of Plato and Aristotle were in harmony with its own view of the world. To the church, Plato's view of the perfect heavens, where the heavenly bodies moved in perfect circular orbits around the Earth, was the only acceptable view of the world. Certainly the Earth, the home of man, created in his own image, was placed by God at the center of his universe. For the church, Ptolemy's universe was the only possible universe.

Humanity would not return to a rational view of the world until the sixteenth century, at the height of the Renaissance, when Galileo invented experimental physics and, with it, the modern scientific method.

It's Only a Theory

Half a century after Copernicus revived Aristarchus's model with the sun at the center of the universe because he was dissatisfied with the inability of Ptolemy's model to explain recent astronomical observations, the Holy Office of the Inquisition called Galileo to testify before them for supporting Copernicus in his latest book:

> Summoned, there appeared personally in Rome at the Palace of the Holy Office . . . Galileo, son of the late Vincenzio Galilei, Florentine, seventy years of age, who, sworn to testify the truth, was asked by the Fathers the following:
>
> . . .
>
> Q: What specifically he discussed with . . . [the] Cardinals [of the Holy Office, in particular with Cardinal Bellarmino].
>
> A: The occasion for discussing with these cardinals was that they wished to be informed of the doctrine of Copernicus, his book being very difficult to understand by those outside the mathematical and astronomical profession. In particular, they wanted to know the arrangement of the celestial orbs under the Copernican hypothesis, how he places the Sun and the center of the planets' orbits, how around the Sun he places next the orbit of Mercury, around the latter that of Venus, then the Moon around the Earth, and around this Mars, Jupiter, and Saturn; and in regard to motion, he makes the Sun stationary at the center and the Earth turn on itself and around the Sun, that is, on itself with the diurnal motion and around the Sun with the annual motion.
>
> . . .

> Q: What decision was made and then notified to him in the month of February 1616.
>
> A: In the month of February 1616, Lord Cardinal Bellarmino told me that since the opinion of Copernicus, taken absolutely, contradicted Holy Scripture, it could not be held or defended, but that it might be taken and used hypothetically.[1]

Galileo's call before the Inquisition was prompted by the publication of his book supporting the Copernican system. But it wasn't just his latest book. Many of Galileo's findings supported the sun-centered system, such as his discovery of the phases of Venus. And Galileo hadn't been shy about using these observations as proof of the validity of Copernicus's model.

Galileo actually sent a copy of his book to the pope expecting him to be convinced by his arguments and to endorse it, settling the issue once and for all. But rather than reading the book, the pope decided to rely on his experts to rule on the issue and decide whether the Copernican system could be condemned as heretical.

The pope called on Cardinal Bellarmino, who personally knew Galileo and who had actually looked through Galileo's telescope. Cardinal Bellarmino had seen the moons of Jupiter that Galileo had recently discovered with his telescope. The cardinal pointed out that the Council of Trent mandated that the interpretation of Scripture must follow the common agreement of the Holy Fathers who knew that the Bible stated very clearly that the sun revolved around the Earth. He wrote that "the words 'the Sun also riseth and the Sun also goeth down, and hasteth to his place where he arouse, etc' were those of Solomon."

Bellarmino told Galileo in no uncertain terms that he could not write about the Copernican system and specifically about the placement of the sun at the center of the universe as fact, but only as a hypothesis because it contradicted the Bible. If he didn't comply, he was clearly told, the Holy Office would take action against him.

Galileo didn't win his battle with the church. But the battle would eventually be won because of Galileo's greatest legacy, his invention of the methods of modern science.

EXPERIMENTING

Galileo's battle with the Aristotelian-Ptolemaic view of the universe to which the church held steadfastly was not prompted by a blind belief in the Copernican system but by a rational, scientific analysis of the two models. Like the Renaissance artists, Galileo's approach to the study of nature was a return to the rational approach of Thales, Anaximander, and Eratosthenes in ancient Greece. But he went beyond them. He realized that if he was going to learn the way nature worked, he would need to move from the passive observation of the phenomena to the careful planning and executing of controlled experiments.

His quest began at the age of twenty-six, when Galileo became a professor of mathematics at the University of Pisa. There he began his studies of mechanics, clashing with Aristotle's ideas about motion, which at the time were widely accepted throughout Europe. The belief then was that heavier objects fell to the ground faster than lighter objects. Galileo had reasoned that this wouldn't be the case and set out to prove it experimentally.

Using "clocks" made with wine bottles filled with water that slowly drained through a small hole in the bottom and marked to indicate the water level, Galileo tried to time falling balls by using a smooth, inclined track to slow down the motion. (His wine bottle clocks weren't fast enough to time a ball falling straight down). After hundreds of experiments, Galileo discovered that all the balls rolled down at the same time, regardless of their mass. He also measured the time it took the balls to traverse the series of marks he'd made on the track and was able to develop a mathematical expression for this motion. With that expression, he could calculate the time it would take an object to fall to the ground from any height and the final speed that it would reach.

Galileo proved mathematically that all objects accelerate to the ground at the same rate, regardless of their mass and that, if dropped from the same height, they would reach the ground simultaneously. His experiments showed that his mathematical equations were correct.

With this approach, Galileo invented the basis of what would become the scientific method, the powerful technique used by scientists to construct theories based on experimental observations and that allows them to make predictions that can be later tested by future experiments. Isaac Newton, his intellectual successor, completed and perfected the method and used it to develop his system of the world.

NEWTON'S CLOCKWORK UNIVERSE

Isaac Newton was born in 1642, the same year that Galileo died.[2] This coincidence highlights the strong connection in the work of these two great scientists. With his extensive studies on the motion of objects, Galileo had discovered that an object in motion would remain in motion forever unless a force acted on it to slow it down or speed it up. Newton used Galileo's remarkable discovery to develop his theory of universal gravitation, the theory that explained for the first time how the universe works on a large scale.

In 1684, Edmond Halley (in whose honor the comet Halley was named) came to see Newton at his home to consult with him on a matter of utmost importance to the scientists at the time: the nature of the force that keeps the planets in its orbits around the sun. Both Halley and the scientist Robert Hooke had recently arrived at the conclusion that this force must be inversely proportional to the square of the planet's distance to the sun, but they couldn't prove it. Halley wanted to know what Newton thought of it:

> After they had been some time together, the doctor [Halley] asked what he thought the curve would be that would be described by the planets supposing the force of attraction towards the Sun to be reciprocal to the square to their distance from it. Sir Isaac replied immediately that it would be an ellipsis. The Doctor, struck with joy and amazement, asked him how he knew it. Why, saith he, I have calculated it.[3]

Newton had discovered the nature of this force twenty years before, when he was twenty-two, having just graduated from college. Although he wanted to continue his studies and start graduate school, he was living on his mother's farm because the great plague had broken out in London and the university was closed. During the eighteen months that he remained at home, his mind unleashed a fabulous output of creativity never seen before and perhaps since. In that year and a half, now known as his *annus mirabilis,* Newton performed experiments to investigate the nature of light and develop his theory of colors, started his astronomical studies of comets, and developed the main ideas that would lead to the invention of calculus. In that year of wonders, he discovered the law of gravity that laid the foundation for his system of the world, explaining how the universe works. "All this was done in the two plague years of 1665 & 1667," wrote Newton later, "for in those days I was in the prime of my age for invention, & minded Mathematics and Philosophy [physics] more that at any time since."[4]

Newton's path toward his discovery of the law of gravity firmly rested on Galileo's studies on motion. As noted, Galileo had discovered that an object in motion and free of any external forces remains forever in motion. He called this motion *uniform.* He also realized that a force altered the steady motion of an object, speeding it up or slowing it down; what he called *acceleration.* Newton formalized these two concepts and cast them in mathematical form in his first two laws of motion. The first law was simply a more formal statement of Galileo's concept of uniform motion: "*Law I.* Every body continues in its state of rest or of uniform motion in a right lane, unless compelled to change that state by forces impressed upon it."[5]

Newton's second law of motion was a refinement and extension of what Galileo had hinted at, that an external force alters the state of uniform motion of a body. Newton expressed it in simple mathematical form and clarified what acceleration meant: a change in the state of motion of an object to speed it up, slow it down, or to change its direction. Newton realized that a change in direction, even if not accompanied by a change in speed, is also an acceleration that requires an external force. Such is the case of an object moving in circles (or ellipses, the paths followed by planets as they move around the sun). "After this time," wrote Newton in the *Principia,* "we do not know in what manner the ancients explained the question, how the planets came to be retained within certain bounds in these free spaces, and to be drawn off from the rectilinear courses, which, left to themselves, they should have pursued, into regular revolutions in curvilinear orbits."[6] And he added, "From the laws of motion, it is most certain that these effects must proceed from the action of some force or other."[7] Newton stated his second law as follows: "*Law II.* The change of motion is proportional to the motive force impressed and is made in the direction of the right line in which that force is impressed."[8]

It was clear to Newton that a force was needed to keep a planet in its orbit about the sun and that such a force was directed toward the sun. But, what was the source of this force? It had to come from the sun itself.

Newton didn't stop there. He proceeded to actually calculate what this force should be. He knew that Johannes Kepler, a contemporary of Galileo, had performed impeccable calculations on the orbits of Mars and had developed three laws that showed how the planets moved around the sun. In particular, Kepler's third law showed a relationship between the time it takes a planet to orbit the sun and its distance to the sun. Combining Kepler's third law with his own second law of motion, Newton was able to show mathematically that the force that the sun exerts on the planets to keep them in their orbits was proportional to the inverse of the square of their distances to the sun. That was the calculation that pleased Halley so much during his visit to Newton in 1684.

Newton's calculation was brilliant. However, the next step required the mind of a genius, the unique mind of Newton. Having obtained the expression for the force with which the sun attracts the planets and keeps them in orbit using Kepler's equations for the planetary orbits, he decided that this same expression should also apply to the force between the Earth and the moon, between the Earth and an apple falling from a tree, and between any two objects anywhere in the universe!

Newton made his equation universal and in doing so he single-handedly explained how the universe works. His universal law of gravitation explained how objects fall to the ground on the Earth, on the moon, or on Pluto; how the Earth and all the other planets revolve around the sun; how the moon orbits the Earth; and how Jupiter's satellites follow their orbits around the giant planet. We know today that his law of gravitation also explains how stars move around the center of the Milky Way galaxy, how galaxies cluster together and revolve around a common center, and how these groups of galaxies form superclusters.

Newton actually tested his great discovery. He knew that if his equation explained how the moon orbits the Earth, then he should be able to use a different method to calculate how much the Earth pulls on the moon due to the force of attraction. Knowing the size of the moon's orbit and the time it takes to revolve around the Earth, Newton made his calculation, obtaining excellent agreement with the results that he had obtained with his law of gravitation.

With his experimental studies on motion, Galileo had developed the rudiments of the modern scientific method. It was left to Newton to complete it. With his calculation to check the validity of his model, Newton did. He started with a mathematical analysis of Kepler's third law of planetary motion, which was based on detailed observations on the motion of Mars. In his mind alone, he developed a mathematical model for how Mars should move due to the force of attraction to the sun. He then extended his model, applying it to all the other planets, to the Earth-moon system, and to any two objects in the universe. Finally, he picked the Earth-moon system and used data obtained by observations to check his model. That pretty much is how scientists work today, and that is what makes science such a powerful tool to discover the secrets of nature.

Newton's universe works like clockwork, moving in predictable ways, directed by his universal law of gravitation. We use Newton's physics today to calculate the interplanetary trajectories of our spacecraft, aiming them to rendezvous with several planets. These trajectories—such as that of the *Cassini* spacecraft to Saturn—use the gravitational attractions of the planets themselves to catapult the spacecraft toward their final destination, an empty spot in space where we have calculated Saturn will be six years, nine months, and seventeen days later (Plate III).

Given ample time and computing power, the equations of Newtonian physics can theoretically give us everything we want to know about the universe. We could, in principle, measure all the relevant parameters in our present universe, input those into Newton's equations, run the clock backward, and obtain the entire history of the universe. We could also run the clock forward and predict the future evolution of the universe. "In the beginning (if there was such a thing)," wrote Einstein in his *Autobiographical Notes,* "God created Newton's laws of motion together with the necessary masses and forces. This is all; everything beyond this follows from the development of appropriate mathematical methods by means of deduction."[9]

In spite of his statement, Einstein gave us a very different universe. It still runs like clockwork, but uses different clocks.

Einstein's Universe

Newton didn't set out to develop a consistent model of the universe. What we call "Newton's universe" is simply the result of a direct application of his laws of mechanics (his three laws of motion and the universal law of gravitation) to the collection of stars that populate the universe. He gave us a universe filled with stars that occupied a location in space, moving according to these laws. Newton thought that the universe was finite in extent while his contemporary and rival Gottfried Leibniz thought that space extended to infinity. For Newton, space was the stage where the stars, planets, and all objects existed and moved. The motion of the bodies in the universe took place not only in this fixed space but also in time. Time was fixed and passed at a constant rate for every observer anywhere in the universe.

In contrast, Einstein deliberately developed a model of the universe based on his recently completed theory of general relativity. He published his model in 1917 in a landmark paper titled, "Cosmological Considerations on the General Theory of Relativity,"[10] a paper that gave birth to a new field in physics, relativistic cosmology. The equations of his general theory led him to propose a universe that was "finite (closed) with respect to its spatial dimensions."[11] Einstein's model was in sharp contrast with Newtonian physics, which seemed to support a universe that extended to infinity and was filled with stars that felt the gravitational attraction of all the other stars, following Newton's laws.[12]

Einstein knew that the model of the universe that had been constructed with Newtonian physics was flawed and proceeded to discuss its contradictions in his paper. He knew that an infinite universe populated with an infinite number of stars was very problematic. Newton's own law of universal gravitation made an infinite, static universe unstable since there would be an infinite amount of gravity pulling at every object and any small disturbance would make it collapse. As Einstein concluded, "The Newtonian stellar system cannot exist at all."[13]

A finite Newtonian universe is also problematic. What happens when you get to the edge of the universe? (Fig. 2–2) Does space simply end abruptly? Is there a boundary and do things, including radiation, bounce off it? If there's no boundary, can you simply stick your hand out and create more universe? What would prevent you from turning on a light beam and letting the photons expand space as they travel through what was empty before?

There are no answers to these questions in Newtonian mechanics. The physics of Newton tells us how the solar system works, how stars interact within galaxies, and how galaxies and clusters of galaxies move. As noted earlier, we use Newtonian mechanics to send our spacecraft to the other planets in the solar system with great success. But Newtonian physics can't tell us how the universe as a whole works or whether it is infinite or has an end.

Figure 2-2. What lies outside the universe? Is there an edge and if so, what does it look like? This illustration is often described as a sixteenth-century woodcut. However, Owen Gingrich of Harvard University thinks that it is likely a piece of art nouveau that was published in *Weltall und Menschheit*, ed. Hans Kraemer (Berlin: Bong, 1907).

A NEW VIEW OF SPACE AND TIME

In his 1917 paper, Einstein mentions the "indirect and bumpy road" that he has traveled to arrive at the first model of the universe—an isotropic, homogeneous, finite, and unbounded static universe. He took his "indirect and bumpy road" for only a few months, probably starting in 1916, right after he finished his general theory of relativity. But the road had began many years earlier, in 1894, when Einstein was fifteen and he asked himself what would happen if he could ride alongside a beam of light. What would he see? Would he be able to see the front of the beam? The question took him on a path that culminated slightly over ten years later in the theory of relativity, a theory that changed our understanding of space and time.

Einstein's resolution of the beam of light paradox required two assumptions about the nature of the universe. These two assumptions, the *principle of relativity* and the *principle of the constant velocity of light*, became the foundations of his theory.

The principle of relativity was an extension of what Galileo had discovered in the seventeenth century: the laws of mechanics are the same for all observers at rest or in uniform motion. "Shut yourself up with some friend in the main cabin below decks on some large ship," wrote Galileo in his *Dialogues concerning the Two Chief World Systems*,

> and have with you there some flies, butterflies, and other small flying animals. Have a large bowl of water with some fish in it; hang up a bottle that empties drop by drop into a wide vessel beneath it. With the ship standing still, observe carefully how the little animals fly with equal speed to all sides of the cabin. The fish swim indifferently in all directions; the drops fall into the vessel beneath; and, in throwing something to your friend, you need throw it no more strongly in one direction than another, the distances being equal; jumping with your feet together, you pass equal spaces in every direction. When you have observed all these things carefully (though there is no doubt that when the ship is standing still everything must happen in this way), have the ship proceed with any speed you like, so long as the motion is uniform and not fluctuating this way and that. You will discover not the least change in all the effects named, nor could you tell from any of them whether the ship was moving or standing still.[14]

Of course you would be able to tell if the ship was moving by looking out through the porthole. That would give you a reference point to discover your motion. In our modern world, it isn't hard to experience this phenomenon firsthand. Sitting on an airplane before pulling out of the gate, we can't tell if we've begun to move unless we look outside. Without a reference point, when you're in the cabin, nothing you do, not even the most sophisticated and precise experiment you could perform, could allow you to discover your motion. According to Galileo, uniform motion and rest are exactly the same thing. The only difference is the reference point. You may be at rest in your chair reading this book, but you're moving around the sun at some thirty kilometers (nineteen miles) a second (kps/mps) along with the Earth. All motion must be related to a reference point. Sitting in your chair, you are at rest relative to the Earth, but you are moving relative to the sun. Clearly the laws of physics don't change because of how you describe yourself sitting on your chair. As Galileo demonstrated with his "thought experiment," the laws of physics are the same whether you are at rest or in uniform motion, regardless of your speed. The same can be said of any other person in uniform motion. From this, Galileo concluded that the laws of mechanics are the same for observers in uniform motion. That's Galileo's principle of relativity.

When Einstein was a physics student at the Polytechnic Institute in Zurich, he became extremely interested in the recent work of James Clerk Maxwell and began to read and study the papers on his new theory of electromagnetism, the branch of physics that Maxwell had just created by marrying and extending electricity and magnetism. Einstein became an expert on electromagnetism, and he soon realized that electromagnetism didn't obey Galileo's principle of relativity. If you were clever enough, you could design a simple experiment in electromagnetism that would allow you to determine if you were moving or not, without looking outside. Einstein considered the consequences of Maxwell's otherwise beautiful and successful theory and decided that it needed to be modified and brought in line with Newton's mechanics. The twenty-six-year-old Einstein proceeded to do just that.

With electromagnetism now obeying the Galilean principle of relativity, Einstein extended the principle to say that *all* the laws of physics are the same for all observers in uniform motion. The principle of relativity led Einstein to discover that the speed of light is constant. According to the principle of relativity, all observers in uniform motion should see no change in the laws of physics. The speed of light plays a fundamental role in nature and Einstein thought that all such observers should measure the same value, that this number should be part of the laws of physics that remain unchanged. He made this discovery the second cornerstone of his theory of relativity.

Einstein's discovery that the speed of light doesn't change for observers in relative motion appears to bring out a contradiction with our understanding of the concept of speed. If you are riding in the back of a subway car that's traveling at 60 miles per hour (mph) and you decide to walk to the front at 3 mph, someone with a radar gun on the ground will clock you at 63 mph. If you decide to return to the back of the car at 2 mph, you'll be clocked at 58 mph. But Einstein's principle says that this isn't the case for light, that the outside observer will measure the same speed for the light from a lamp in the subway car regardless of whether the car is standing still at the station or moving at 60 mph.

Einstein was conflicted with this problem for some time during his development of the theory of relativity. The formal development of the theory took Einstein all of five weeks in 1905. At the time, he was an unknown clerk at a patent office in Bern, Switzerland, having failed to obtain an academic job because he'd antagonized his professors with his rebellious character and, as a result, no one would give him a good letter of recommendation. During those years, Einstein's lifelong friend Michele Besso would accompany him on their daily walks to and from work discussing Einstein's latest research. During one of the evening walks, Einstein brought up his conflict with the nature of the speed of light and the speeds of ordinary objects. Einstein stayed over at Besso's late into the night discussing this issue and went home without having resolved the problem. However, during the night, the solution came to Einstein. The next morning he greeted Besso with the news that he'd solved the problem.

"My solution was . . . for the very concept of time," he wrote later.[15] That evening, Einstein realized that time, like space, was relative. Space and time aren't fixed, like Newton thought, but change as the observer moves. But more important, space and time change in such a way as to keep the speed of light constant. Space is relative, time is relative, but the speed of light is absolute; it remains always the same, regardless of the motion of the observer.

With his theory of relativity Einstein finally resolved the paradox he'd encountered at the age of sixteen. Light always travels at the same speed of 300,000 kps (188,000 mps) for any observer in uniform motion. Regardless of how fast you move, you'll never see the front of the wave, because for you, light moves at the same 300,000 kps.

More important, Einstein's theory of relativity gave us a new meaning for space and time. Space is no longer the stage where things take place and time no longer flows at the same rate for everyone, as they were for Newton. Space and time are relative, changing as you move, while the speed of light remains unchanged.

The changes in space and in the flow of time for observers in motion relative to one another are extremely small, requiring extremely precise and sophisticated measurement techniques. That's why no one had noticed the phenomenon before. Only years after Einstein proposed his theory were the two phenomena measured experimentally.

A More General Relativity

Soon after the publication of his theory of relativity, Einstein started to think about its limitations. His theory, later called the Special Theory of Relativity, dealt only with uniform motion—motion along a straight line at a constant speed. It excluded accelerated motion, when an object speeds up, slows down, or turns. The reason was that the principle of relativity on which the special theory is based applies to motion that cannot be detected, and accelerated motion can always be detected. You're always able to detect when you're riding in a bus or subway if the vehicle changes speed or turns.

If relativity is based on the principle that uniform motion is undetectable and accelerated motion is detectable, how can it be included in the theory? Einstein decided that he needed another property that would remain undetectable. One day, while sitting in the patent office in Bern, it suddenly occurred to him that a person falling from the roof of his house would not feel gravity. That was the "happiest thought of my life," he said during a lecture on relativity that he gave years later in Japan.[16]

With his happy thought, Einstein found the property that remained undetectable. When you're falling, you don't feel gravity; that is, you're weightless. If you're weightless, you could be falling or you could be in interplanetary space. If you close your eyes, you couldn't tell the difference. In fact, there's no experiment you could do that would allow you to discover which is happening. When you're falling, you're accelerating toward the ground due to the gravitational attraction of the Earth. When you're in interplanetary space, you're far away from any celestial body that could exert a measurable gravitational attraction on you.

In a comprehensive paper on the special theory published in 1907,[17] Einstein used a thought experiment to illustrate his idea. Imagine a laboratory aboard an interplanetary spaceship that's accelerating at 1 *g* (the same as the gravitational acceleration near the Earth's surface). In this accelerating laboratory, a scientist lets go of a ball he's been holding. After it is released, the ball no longer accelerates and continues moving with the speed that it had at the moment of its release. Because the ship is accelerating at 1 *g*, its speed increases continuously and it catches up with the ball, which is simply cruising. The scientist measures the speed and location of the ball at several time intervals and records the data. The following day, the scientist walks to his lab and continues his measurements, noticing that they don't deviate from what he'd taken the day before. When he looks out the window of the spacecraft, however, he realizes that while he slept the spacecraft had landed back on Earth, and the ball is falling to the floor because of the gravitational attraction of the Earth and not because of the acceleration of the ship.

Einstein concluded that acceleration and the effects of gravity cannot be distinguished; they are the same phenomenon. This idea, which he called the *principle of equivalence*, became the foundation for his extension to the special theory of relativity, his *general theory of relativity*.

It took Einstein about eight years of extremely hard work to fully develop his general theory of relativity, which is considered Einstein's masterpiece. Shortly after its publication in 1915, Einstein became a celebrity due to the striking confirmation of one of the strangest predictions that he made with the theory, the phenomenon of light deflection by gravity. The spaceship in Einstein's thought experiment can illustrate the prediction. Imagine that one of the scientists on the spaceship accelerating at 1 *g* has installed a tiny window on the side of the ship to examine a narrow light beam from a distant star (Fig. 2–3). While the light beam traverses their lab, the ship accelerates and the beam hits the opposite wall at a point not in line with the small window and the star. The scientist observes that, from her perspective, the light beam has bent downward. Seen from outside the ship by an observer not accelerating, the light beam has continued its straight path while the ship had accelerated forward.

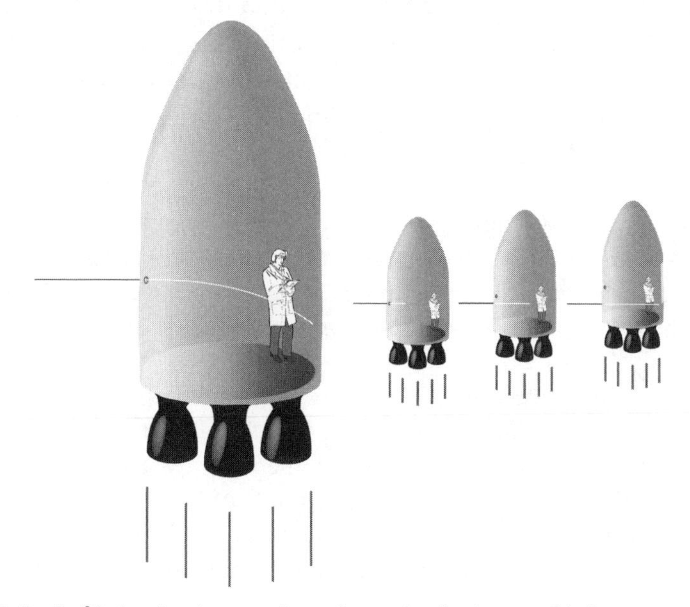

Figure 2-3. (Left) A scientist examines the path of a beam of light entering the lab aboard her spaceship accelerating at 1 *g*. From her perspective, the beam bends downward. (Right) As seen from the point of view of a nonaccelerating observer outside the ship, the beam continues moving along a straight line, while the ship accelerates forward.

Notice that if the ship had not been accelerating but simply moving at a constant speed relative to an outside observer on a moon base, for example, the scientist on the spaceship would have observed the light beam also striking the opposite wall at a lower point than expected, but the path of the light beam from the small window to the wall would have been a *straight* line, not a curve. It's only when the spaceship accelerates that the path is a curve, since the ship's speed is continuously increasing, traveling larger and larger distances as the light beam crosses the ship at a constant speed.

According to the principle of equivalence, the ship's 1 *g* acceleration is the same as the Earth's gravity. Consequently, the gravitational attraction of the Earth should bend light. Einstein's calculations told him that this bending was too small to be detected even with the most precise instrumentation. What was needed was a much stronger gravitational attraction, such as that from the sun. The starlight could be seen during a solar eclipse, when the moon blocks out the disc of the sun. In 1919, the British astronomer Arthur Eddington, a member of the Royal Society, organized a special expedition to the island of Principe in West Africa to check Einstein's prediction. In his book *Space, Time, and Gravitation*, Eddington wrote:

> On the day of the eclipse the weather was unfavorable. When totality began the dark disc of the moon surrounded by the corona was visible through cloud, much as the moon often appears through cloud on a night when no stars can be seen. . . .
>
> There is a marvelous spectacle above, and, as the photographs afterwards revealed, a wonderful prominence-flame is posed a hundred thousand miles above the surface of the sun. We have no time to snatch a glance at it. We are conscious only of the weird half-light of the landscape and hush of nature, broken by the calls of the observers, and beat of the metronome ticking out the 302 seconds of totality. . . .
>
> [Of the sixteen photographs obtained, only] one was found showing fairly good images of five stars, which were suitable for determination. . . . The results from this plate gave a definite displacement, in good accordance with Einstein's theory and disagreeing with the Newtonian prediction.[18]

Einstein's prediction was correct: Light is deflected by the gravitational attraction of the sun!

Einstein's completed general theory of relativity appeared in the January 1916 issue of the journal *Annalen der Physik* with the title "The Formulation of the General Theory of Relativity." The new theory extended the relativity of motion to include all types of motion, uniform and accelerated. It also explained gravity, equating it to acceleration. The general theory is then a theory of gravity, an extension to Newton's mechanics.

Warping Spacetime

The main consequence of the special theory of relativity is that space and time aren't fixed but change with the motion of the observer. Moreover, space and time are entangled through the speed of light and, for Einstein, the two concepts, space and time, were on an equal footing. When you stand still relative to the Earth, you have no motion through space but you're moving through time. When you begin to move relative to the Earth, part of your motion through time changes into motion through space. If you increase your speed, more of your motion through time gets converted into motion through space. In a sense, you move less through time so that you can move more through space. The limiting case is the speed of light, when all the motion is through space and none through time. Einstein showed that this limiting velocity couldn't be reached by any massive object. However, photons, the massless carriers of light, don't move through time; time doesn't flow for them.

Since space and time are linked and change into each other, Einstein found it necessary to consider a single four-dimensional *spacetime.* This four-dimensional spacetime contains the three dimensions of space and one of time.

In his formulation of the general theory of relativity, Einstein needed to introduce a curved spacetime to describe the gravitational field. This spacetime becomes distorted or curved by the presence of a gravitational field; that is, by the presence of a body. A common and useful analogy to illustrate the curvature is to imagine the four-dimensional spacetime as being represented by a two-dimensional elastic membrane, like a trampoline (Fig. 2-4). The undisturbed, stretched membrane represents spacetime in a region devoid of any masses, far away from any gravitational fields (Fig. 2-4 Left). A small, light marble will roll on the trampoline along a straight line, creating a negligible depression on it. However, a heavier bowling ball placed in the middle of the trampoline makes a dip and the marble will roll toward it. If you roll the marble on the trampoline, it will follow a curved path around the dip (Fig. 2-4 Right). The bowling ball isn't pulling the marble; it distorts the trampoline and this distortion affects the motion of the marble. Similarly, the sun creates a dip in spacetime, distorting it, affecting the motion of the Earth and the other planets in the solar system. In general relativity, gravity is not a force somehow generated by the sun and acting instantaneously on the Earth and the planets. Gravity is the distortion in spacetime caused by the presence of the sun. For Einstein, gravity is the geometry of spacetime. The physicist John Wheeler put it in simple terms: "Matter tells space how to bend, space tells matter how to move."[19]

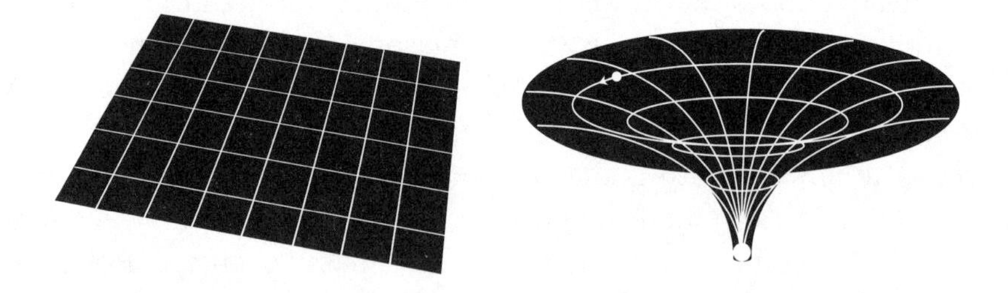

Figure 2-4. (Left) A stretched elastic membrane, like a trampoline, is a good two-dimensional representation of the four-dimensional spacetime that Einstein proposed. A flat, undisturbed membrane represents the spacetime in a region of space with no gravitational fields present. Motion of a small mass in this undistorted spacetime takes place along a straight line. (Right) A small mass moving in the distorted spacetime created by a large mass represents a planet moving in the distorted spacetime around the sun.

The distortion of spacetime explains the deflection of light by the sun. Light travels through spacetime along a straight line. Since spacetime is distorted, a *straight* line is the shortest path between two points—a geodesic—much like the shortest path between two distant points on Earth is an arc on the spherical surface of the planet. Since the sun curves or distorts spacetime, light bends to follow the geometry of spacetime.

Einstein's System of the World

General relativity is a theory of gravity, an extension of Newton's own theory of gravity. Like Newton's mechanics, Einstein's general relativity is a system of the world. Newton's universal law of gravitation is a straightforward force that acts instantaneously between any two objects anywhere in the universe. The force acts through space, but doesn't interact with space. Space, for Newton, exists only as a stage for matter to exist in and move through but doesn't have material existence and doesn't affect matter. For Einstein, space and time aren't fixed; they change with the motion of the observer. Space and time are also linked to each other insofar as an increase in motion through space is accompanied by a corresponding decrease in motion through time. This interconnection between the two means that they are actually two parts of a whole, parts of spacetime. In Einstein's theory, matter interacts with spacetime, disturbing it by its presence. Conversely, spacetime interacts with matter, constraining its motion. The distortion in spacetime is what we call gravity.

In his 1917 cosmology paper, Einstein applied his equations of general relativity to the entire universe and was able to develop a consistent mathematical model of the universe. Spacetime is distorted around each one of the innumerable stars encountered throughout the universe. As scientists found out later, these stars are concentrated in galaxies, island universes like our own Milky Way, which contain 100 billion stars (Plate IV). The distribution of stars in the universe varies from the large density of stars concentrated in a galaxy to zero stars in the region between galaxies, and the curvature of spacetime varies according to this fluctuation. On a large scale, however, the distribution of galaxies in the universe is fairly close to being uniform and the smoothed-out overall density of the universe is about one hydrogen atom per cubic meter. As a result, the universe as a whole has a smooth curvature throughout.

Einstein's model extended the local curvature of space generated by matter to the entire universe. The universe that emerged from general relativity was finite but unbounded, with stars and galaxies curving space. The three dimensions of space don't extend out to infinity; rather, they are closed, in the same way that the two dimensions of a sphere are closed.

In a popular exposition of his theories Einstein illustrated his finite and unbounded universe by having us imagine flat beings existing in a two-dimensional sphere. "The flat beings with their measuring rods and other objects fit exactly on this surface and they are unable to leave it. Their whole universe of observation extends exclusively over the surface of the sphere."[20] For these flat beings, nothing exists outside the surface of their spherical universe. These beings can travel throughout their entire universe, never reaching an edge. If they travel for a long time along a straight line, they would eventually come back to their starting point, having traveled all the way around the universe. As Einstein wrote in his book, "The great charm resulting from this consideration lies in the recognition of the fact that *the universe of these beings is finite and yet has no limits.*"

Like the spherical world of the flat beings, the three-dimensional space of our universe is finite in extension but unbounded, so that there is no edge, no boundary for you to stick your hand out through. The question of what lies outside the universe, which can be posed to a Newtonian universe, has no meaning in Einstein's model, since there is no outside.

Einstein's general relativity equations initially gave him a dynamic universe, one that was collapsing or even expanding, not the static universe that the astronomers observed. Analyzing his equations carefully, he concluded that there was no solution that would describe a static universe. But he persevered. His equations allowed for some flexibility, and he took advantage of it. "The conclusion I shall arrive at," wrote Einstein in his 1917 paper, "is that the field equations of gravitation, which I have championed hitherto, still need a slight modification, so that on the basis of the general theory of relativity those fundamental difficulties may be avoided."[21] He modified his equations to fix the instability that would cause the collapse or the expansion by adding a term, the *cosmological constant*, as he called it.

Einstein was uneasy about taking that step. What he'd just done was not characteristic of him. What Einstein always did was to stick with his equations regardless of where they took him. Not this time. Soon, events would tell him whether he was right in his decision.

Chapter 3

IN THE BEGINNING

THE GEOMETRY OF THE UNIVERSE

According to Einstein himself, his introduction of the cosmological term was the greatest blunder of his life.[1] Was he right in making this rather strong assessment about his own work? In his 1917 cosmology paper he concluded that his model of the universe was logically consistent and that to achieve this consistency he "had to introduce an extension of the field equations of gravitation which is not justified by our actual knowledge of gravitation."[2]

Einstein's cosmological term counterbalanced the gravitational attraction, being in effect an antigravity force. However, the structure of the field equations was such that this added antigravity term had negligible effect at short distances, becoming meaningful only at very large distances. General relativity's extremely successful predictions of the planetary orbits (in particular, the peculiar orbit of Mercury), unexplained until Einstein applied his equations to this case, were not affected by the cosmological antigravity term.

Einstein's model agreed with the experimental evidence and was well accepted by the cosmologists at the time. But not by all. The Russian scientist Alexander Friedmann studied Einstein's paper and decided to remove the cosmological term and return to Einstein's original form of the field equations to see where they would take him. What Friedmann found was that the original equations did produce a dynamic universe, as Einstein had predicted. However, unlike Einstein's conclusion that the universe was unstable, Friedmann's analysis gave him an expanding universe, not an unstable one, as Einstein had assumed.

Alexander Friedmann was probably one of a handful of scientists who could have tackled Einstein's model and modify it. He was working in St. Petersburg, Russia, where the First World War, the subsequent Bolshevik Revolution, and the civil war that ensued had made the scientific literature unavailable. This relative isolation from mainstream physics may have given Friedmann the opportunity to look at things from a different perspective.

Friedmann was born in 1888 in St. Petersburg and studied mathematics at the university in his hometown. There he attended the seminars that the physicist Paul Ehrenfest had organized on the quantum theory being developed at the time, and on relativity and statistical mechanics. After completing his graduate studies, he obtained a position in the Aerological Observatory in St. Petersburg, participating in several observational flights. His work stimulated an interest in aviation, and during the First World War he joined the aviation detachment, participating in several bombing missions. He used his knowledge of physics to predict the bomb trajectories, writing to a friend that he'd been able to verify the predictions of his calculations of where the bombs should hit.

After the war, he was appointed director of the Central Aeronautical Station in Kiev. A year later, the Bolshevik Revolution of 1917 closed the station and Friedmann went to the University of Perm, where he started his academic career as a professor in the department of mathematics and physics.

By 1920, Friedmann moved to the University of St. Petersburg, where he became interested in Einstein's theory of general relativity, which had been published in 1915 but had not been known in Russia due to the wars. In 1922, after a careful study of Einstein's 1917 cosmology paper, Friedmann discovered that Einstein's equations in their original form allowed for the possibility of two dynamic models. One described an expanding universe and the other a contracting universe.

According to general relativity, each star curves spacetime by an amount that is related to the star's mass. The curvature of the entire universe depends on its total mass or, equivalently, on its mass density (the average mass per unit volume). The greater the mass density of the universe, the greater its overall curvature. Friedmann found that the two dynamic models allowed by Einstein's cosmological equations were related to the mass density. If the universe has a very large mass density, it would first expand but the expansion would continuously slow down due the large gravitational force of all the stars and galaxies. The expansion would eventually stop, and the gravitational attraction would pull the universe back into a contraction. This type of universe would have spherical curvature and is called a *closed universe* (Fig. 3–1, Left).

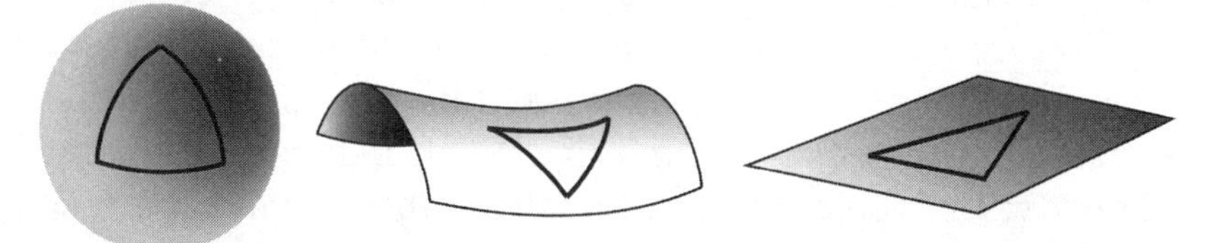

Figure 3-1. The three types of possible dynamic universes. (Left) A *closed universe,* with spherical geometry. The interior angles of a triangle in this universe add up to more than 180°. (Middle) An *open universe* has a geometry that curves away from itself, like the surface of a saddle. A triangle in this universe has its interior angles add up to less than 180°. (Right) A *flat universe* has no curvature. The interior angles of a triangle add up to exactly 180°.

If the universe has a very low mass density, the gravitational attraction of all the objects in the universe would not be strong enough to stop the expansion. The universe would continue expanding at the same rate forever. The curvature of this universe would not be closed, like the spherical curvature of the first type, but open, like the surface of a saddle (Fig. 3–1, Middle). Fittingly, this type is called an *open universe.*

At the boundary between the open and closed universes, the mass density of the universe has a *critical value.* If the universe happens to have exactly this critical mass density, the gravitational force that slows the expansion would not be large enough to stop it completely. In this case, the universe would expand forever. At this boundary, there would be no curvature, and the universe is *flat* (Fig. 3–1, Right).

Friedmann didn't consider the possibility of a flat universe. Perhaps it was too much to expect that the mass density of the universe would have exactly the critical value. For him, the universe was expanding either as an open or as a closed universe.

Friedmann was the first scientist to advance the idea of a universe changing in time and evolving. He published the results of his research in 1922 and 1924, but his work remained largely unnoticed. Einstein did notice it and initially disagreed with this model, thinking that it was incorrect. After a more careful examination, Einstein acknowledged that the model was after all correct. In his *The Meaning of Relativity*, Einstein wrote: "[Friedmann] showed that it is possible, according to the field equations, to have a finite density in the whole (three-dimensional) space, without enlarging these field equations *ad hoc*."[3] However, Einstein didn't think that the new model represented the real universe and Friedmann himself had to accept that the observations didn't support his model. "All this should at present be considered as curious facts," he wrote, "which cannot be reliably supported by the inadequate astronomical experimental material."[4]

In spite of his cautious statement, Friedmann felt that his model did describe the actual universe and felt victorious when Einstein finally conceded that his model at least was correct. For Friedmann, life was beginning to turn for the better. To top it off, he and his wife were expecting their first child.

But the good times weren't going to last. In July of 1925, just a year after the publication of his second paper on his model, Friedmann contracted typhoid fever and died within a few days. He was thirty-seven years old.

REINVENTING FRIEDMANN'S MODEL

At the time of Friedmann's death, a young new instructor of physics was starting his academic career, working on the new cosmology based on Einstein's general relativity. George Lamaître had finished the coursework for his doctorate at the Massachusetts Institute of Technology and had returned to his native Belgium to take up a position at the Catholic University of Louvain. During the next two years, Lamaître worked on his doctoral thesis in the field of general relativity. At the same time, he developed a dynamic model for closed universes based on Einstein's field equations. That was, of course, what Friedmann had done. Although Lamaître quoted Friedmann's 1924 paper, he apparently was not aware of the details of that work. He probably obtained the reference from another paper without actually reading Friedmann's original publication, which hadn't appeared in a mainstream journal.

Lamaître went in a different direction than Friedmann, however. Friedmann had done away with Einstein's cosmological constant and had proposed two possible models, one for a closed universe and a second one for an open universe. Lamaître, on the other hand, kept the cosmological constant and proposed only a closed universe. His model was very similar to Einstein's. Einstein's model had a cosmological constant that balanced the universe exactly at the unstable point between an open and a closed universe. Lamaître's value for the cosmological constant simply let the universe expand.

While Friedmann interpreted his solutions to Einstein's field equations as describing either the expanding or the contracting phases of the universe, Lamaître interpreted the dynamic, expanding universe of Einstein's equations as evidence for the universe's origins. If the universe was expanding now, it must have been much smaller in the past. At some point in the distant past, all the matter in the universe must have been compressed into an extremely dense and hot state of matter concentrated in a tiny volume; he called this volume the *primeval atom*. At the instant of creation, this superdense primeval atom disintegrated—much like the disintegration of radioactive atoms into smaller atoms—starting a very rapid expansion, thinning out and cooling the matter, which coalesced into the stars that we see today. "The primeval atom hypothesis is a cosmogenic hypothesis," he wrote, "which pictures the present universe as the result of the radioactive disintegration of an atom."[5]

Lamaître's primeval atom hypothesis, along with Friedmann's dynamic model of the universe, are the early seeds of the big bang, the modern theory of the evolution of the universe. Lamaître published his primeval atom hypothesis in 1927, a couple of years after Friedmann's second and last paper on his model. At these early stages, however, Lamaître and Friedmann's ideas were simply not supported by observations. To most astronomers the universe was static and, except for local changes, had existed in the current state forever. Two years later, all of this would change.

THE EXPANSION OF THE UNIVERSE

In 1914, a year before Einstein completed his general theory of relativity and before he even started thinking about building a model of the entire universe, the American astronomer Vesto Slipher at the Lowell Observatory in Arizona made an interesting discovery. He'd been studying distant clusters of stars for several years and noticed that the spectra of many of them was Doppler-shifted toward the red, indicating that they were moving away from us.[6]

Slipher was making his observations in the middle of World War I, when communications and the availability of scientific literature were restricted. As a result, Einstein and Friedmann were unaware of Slipher's work.

In 1924, another American astronomer by the name of Edwin Hubble discovered that many of the clusters of stars that Slipher and others had been observing were actually *island universes*, vast collections of stars external to our own Milky Way. Using the recently inaugurated one-hundred-inch telescope at Mount Wilson Observatory in California, Hubble was able to resolve individual stars in those clusters. His discovery of *galaxies* that contain billions of stars is considered among the greatest discoveries in astronomy (Plate V).

Edwin Hubble was not a man easy to befriend. He was a brilliant scientist, eventually becoming the foremost astronomer of his time. He also was somewhat pretentious and affected. When he was at the University of Chicago he studied law to appease his strong-willed father, but he also studied physics and astronomy. Earning a Rhodes scholarship, he spent two years at Oxford, where he transformed himself into a fake Englishman, speaking with a British accent. Warren Ault, a historian and fellow Rhodes scholar, found him toward the end of his stay at Oxford dressed in a Norfolk jacket and cap, using a cane, and speaking with such a thick accent that he had trouble understanding him.[7]

His extreme and rapid transformation is evident in a letter that he wrote to his mother a few days after his arrival in England. Hubble wrote that several of his fellow Rhodes scholars were "splendid fellows," and that even Edwards of New Zealand, although not a Rhodes man, still was "a mighty good sort." Edwards had lent him his "wheel," but "as soon as I get on sound financial footing," he told his mother, "I shall straightway purchase one of my own."[8] The other Rhodes scholars were laughing behind his back at his effort to speak with an extreme English accent. "We always claimed that he could not be consistent," wrote one classmate, "so that he might take a bäth in a băth tub."[9]

At Oxford, Hubble studied law, and upon his return to the United States, passed the bar exam. He practiced law for a year but soon decided that his heart was in astronomy and returned to the University of Chicago, obtaining his doctorate in 1917. He then set out to obtain a position at Mount Wilson, the observatory with the largest telescopes in the world. However, when the offer came, the United States had entered World War I, and Hubble decided that he needed to enlist so that he could help defend his beloved Britain. After the war, Hubble returned to America, joining Mount Wilson Observatory in 1919. The astronomer from Marshfield, Missouri, kept his adopted British accent throughout his life.

Although hard to befriend, except to a few, Hubble was respected and admired for his work. His assistant and collaborator Milton Humason described Hubble during an observation night at Mount Wilson, when the conditions were far less than ideal:

> 'Seeing' that night was rated as extremely poor on our Mount Wilson scale, but when Hubble came back from developing his plate in the darkroom he was jubilant. 'If this is a sample of poor seeing conditions,' he said, 'I shall always be able to get usable photographs with the Mount Wilson instruments.' The confidence and enthusiasm which he showed on that night were typical of the way he approached all his problems. He was sure of himself—of what he wanted to do, and of how to do it.[10]

After his important discovery of the existence of galaxies, Hubble started measuring their distances and studying their properties in an attempt at a classification scheme. By 1929, he had measured the distances to twenty-four galaxies. When he compared these distances to the recession velocities of the galaxies calculated from the spectra that Vesto Slipher had measured, he discovered that all galaxies were moving away from each other and that the farther they were, the faster they moved. The recession velocities of distant galaxies were proportional to their distances. Hubble discovered that the universe is expanding!

Hubble had made the most important discovery on the nature of the universe. He provided experimental proof for Friedmann and Lamaître's dynamic universe models. Lamaître's reasoning that an expanding universe implies a beginning, an origin, a moment of creation, had some validity now.

Hubble's paper announcing the results of his discovery was published in March 1929.[11] Einstein, who had only agreed to the correctness of Friedmann's model but didn't think that it represented the real universe, had to give in and accept that he'd made his greatest mistake. Friedmann didn't live to see the experimental verification of his model. Lamaître, on the other hand, enjoyed worldwide recognition. In his 1927 paper where he proposed his primeval atom hypothesis, he had predicted that the galaxies should be moving away from each other with speeds proportional to their distances. That was exactly what Hubble's observations of the motion of galaxies had revealed.

The relationship between the recession velocities of distant galaxies and their distances is now known as the *Hubble law*. This significant expression gives the expansion rate of the universe, or *Hubble constant*. In his 1929 paper, Hubble calculated this expansion rate from his data and came up with the value of 150 kilometers per second per million light-years of distance. This value turned out to be too high by a factor of five to ten!

The reason for Hubble's incorrect value in his original calculation was that his measurements of the galactic distances were inaccurate. There were several methods to measure these distances when Hubble was undertaking his landmark study. The uncertainty of these methods became worse for large distances. Since Hubble's law is very sensitive to inaccuracies at either distance extreme, short and large, his initial result was way off. Modern measurements place the value of the Hubble constant at 15–30 kilometers per second per million light-years.

Since the Hubble constant relates the recession velocity of a galaxy and its distance, its value provides knowledge of the age of the universe. This is similar to your ability to figure out how long you've been traveling by knowing the speed and the distance traveled. If you travel 200 kilometers by train at a steady speed of 100 kilometers per hour, you know you've been traveling for 2 hours.[12] Hubble's original value set the age of the universe at only two billion years, much shorter than the age of the Earth obtained from geological data.

In his 1927 paper, Lamaître used an early value of Hubble's constant, which he likely obtained directly from Hubble or from his collaborators. This value, 190 kilometers per second per million light-years, was much higher than the incorrect value that Hubble would publish two years later, and made things ever worse for Lamaître's idea of the primeval atom disintegrating and originating the expansion. Taking a cue from Einstein, Lamaître decided to use the cosmological constant in his model to adjust the age of the universe to a value much larger than the value obtained directly from his chosen Hubble constant.

Lamaître's model has only a historical value today. His main contribution to cosmology was in the introduction of the idea of the origin and expansion of the universe, what would later on evolve to become the big bang theory.

Making the Stuff of Matter

Einstein's general theory of relativity made possible the development of the general ideas of the big bang theory. His own model and the modifications of Friedmann and Lamaître, along with the observational proof of the expansion of the universe provided by Hubble, showed us that theoretical physics together with observational astronomy could one day uncover for us the nature of the universe.

But difficulties remained. Hubble's early estimate of the rate of expansion yielded a universe younger than the Earth. Soon, other difficulties began to emerge. The main one had to do with the formation of the elements out of a primeval atom. Astronomers knew that most of the universe, 99.9 percent of it in fact, was composed of hydrogen and helium with very small percentages of the heavier elements. The most abundant of the heavy elements was oxygen, followed by carbon. These are modern measurements of their relative abundance:

Element	Relative Abundance in %	% of Universe Mass
Hydrogen	90.9	73
Helium	9	25
Oxygen	0.06	< 1
Carbon	0.02	0.5
All others	0.01	0.5

No known mechanism could explain this specific relative abundance. It was time to reconsider the formation of the early universe.

George Gamow began to think about the early universe in new ways. As noted earlier, the universe at the present point in its expansion has a density that astronomers have calculated to be about one atom per cubic meter. As you go back in time, the universe compresses to a higher density. As happens with a gas, this compression would cause the temperature of the universe to increase. Gamow knew that astrophysicists had been able to decipher the mechanism of energy generation in stars from their knowledge of the temperature and density conditions inside the sun. He decided to use similar techniques to figure out the conditions of the very early universe.

Gamow, then a professor of physics at George Washington University, had a good pedigree. He had learned relativity from Friedmann when he was a student at the University of St. Petersburg in the late 1920s, right after Friedmann published his papers. Born in the Ukraine, George Gamow immigrated to the Unites States in 1934. However, his emigration from the Soviet Union was not uneventful. Unhappy with the Soviet regime, he attempted to flee with his wife in a kayak across the Black Sea. The two began paddling in 1931 from the Crimean peninsula, heading toward Turkey some 250 kilometers (150 miles) away. After a day and a half of paddling, a storm interrupted their audacious attempt and they were forced to return. Unfazed, Gamow tried again, attempting to cross the Artic toward Norway. Just like the first time, this second attempt failed.

A few years earlier, Gamow had been a fellow at the Copenhagen Institute for Theoretical Physics, where he had met Niels Bohr. Hearing of his old friend's difficulties in leaving the Soviet Union, Bohr intervened with the help of the physicist Paul Langevin, who had communist connections. Langevin was able to convince the Soviets to appoint Gamow as a delegate to the Solvay Conference to be held in Brussels. After long negotiations with the Soviet authorities, Gamow was able to obtain a passport for his physicist wife as well. Unaware of Bohr's involvement, Gamow and his wife didn't return to the Soviet Union after the conference ended. A few days later, they learned of Bohr's negotiations. They also learned that Bohr had misunderstood their intentions and had given Langevin assurance that Gamow would be returning to Russia at the end of the conference. Eventually Marie Curie smoothed things out with Langevin and Bohr so that Gamow and his wife could remain in Europe.

Gamow is usually considered to be the originator of the big bang theory. Although Lamaître did come up with the general idea for the expansion from a primeval atom, it was Gamow who first uncovered the actual mechanism and the physics of the very early universe. In 1946 he proposed that the early universe was a hot gas of neutrons. Gamow knew that neutrons are unstable and decay into protons and electrons in a process known as *beta decay*.[13] According to Gamow's model, a second nuclear reaction that generates neutrons from these particles replenished the neutrons lost in the beta decay process, establishing an equilibrium. During the first moments of the expansion, the universe must have consisted solely of neutrons, protons, and electrons, since, at the extremely high temperatures of the young universe, any particles that combined into a nucleus would have dissociated immediately. Gamow called the initial collection of particles the *ylem*, from a Middle English word meaning "the first substance from which the elements were supposed to be formed."

In Gamow's model, when the density and temperature began to drop as a result of the rapid expansion of the early universe, the energy-dependent, neutron-producing reaction slowed down and eventually stopped. According to Gamow's calculations, five minutes after the expansion started, the universe had reached low enough temperatures to allow some of the particles in the ylem to stick together, forming atomic nuclei. First, a neutron and a proton came together to form the nucleus of what we know today as deuterium, a form of hydrogen. Then a nucleus of deuterium joined with a neutron to form the nucleus of tritium, a third form of hydrogen. Subsequent thermonuclear reactions by successive neutron capture would form all the nuclei that are present in the universe today. The entire *nucleosynthesis process*—as he called it—would have ended after thirty minutes, since Gamow's calculations showed that, by then, the expanding universe would have reached temperatures too low for thermonuclear reactions to take place. Moreover, since the lifetime of neutrons is only about twelve minutes, most of the neutrons that hadn't been used in forming the elements would have decayed into protons and electrons already.

Gamow and his graduate student Ralph Alpher spent three years attempting to reconstruct theoretically the sequence of events that they thought took place in the early moments of the universe. Together, they were able to model the formation of two additional nuclei: helium-3, consisting of two protons and one neutron, and helium-4, with two protons and two neutrons. Their model showed that at the end of the nucleosynthesis phase, there were ten nuclei of hydrogen for every nucleus of helium, the ratio that astronomers observe today.

Encouraged by this success, Gamow and Alpher plowed on. But try as they might, they were not able to move forward. The problem that stumped them was that a nucleus with five *nucleons* (the collective name for protons and neutrons) does not exist. The peculiar and already known way in which the nuclear forces act on the nucleons forbids the existence of this nucleus. To continue modeling the formation of the elements, they had to bypass the 5-nucleon island, but that proved impossible. They tried all possible avenues and discarded them as being extremely unlikely or not allowed, according to the known properties of the nuclei involved.

Gamow and Alpher reported their findings in a paper titled "The Origin of Chemical Elements," which they planned to submit to the prestigious journal *Physical Review*. Gamow, who was always fun loving and adventurous, decided that since their two last names, Alpher and Gamow, sounded like alpha (α) and gamma (γ), he needed to add his friend Hans Bethe (whose last name is pronounced "beta" [β]) to the list of authors. Bethe, who had discovered the mechanism for energy production of stars for which he would later win the Nobel Prize, had not contributed anything to this work, but agreed to his friend's unusual request, unaware that Alpher was unhappy about the whole affair. Since Alpher was the unknown student, he said nothing to his advisor and the "α β γ" paper was published, by coincidence, on April 1, 1948.[14]

That same year, Gamow published a paper[15] showing that when the nucleosynthesis phase ended, the universe had cooled down to a million degrees. At those temperatures, the newly formed nuclei would not pair up with the free electrons to form atoms and all the matter in the universe would be a plasma, consisting of nuclei and free electrons. There would also be photons scattering off the electrons, just as light scatters off water droplets in a cloud. And just as we can't see through a cloud, we wouldn't have been able to see through the fog of the scattered photons in the early universe. At this stage, the universe was opaque to light.

Alpher, who had graduated now, began a collaboration with Robert Herman and the two decided to work on expanding the details in the life of the early universe, following Gamow's latest paper. According to their calculations, the plasma phase of the universe lasted for 300,000 years, by the end of which the universe had cooled down to 3,000°C (5,400°F). At that point the universe entered the phase of the atom. Since atoms are electrically neutral, the photons, which were interacting only with charged particles, would have begun to uncouple. The universe was now transparent to light. This event is called *recombination*, which is a bit of a misnomer since at this moment the nuclei and electrons began to combine for the first time.

During the opaque phase of the universe, the multiple scattering of the photons produced what physicists call a *blackbody*, or *thermal spectrum*. A spectrum is the variation of the energy of the radiation with the frequency or the wavelength of the light. A blackbody spectrum, which occurs in several other situations in physics, is the spectrum of a radiating body in equilibrium and is independent of the material of the body. The blackbody spectrum has a universal form, as seen in Fig. 3–2. In the case of the big bang, the photons would reach a steady-state condition or equilibrium due to the multiple interactions with the charged matter in the plasma.

Alpher and Herman realized that these original photons, which could no longer interact with the neutral atoms forming for the first time throughout the universe, would have survived and would exist today as background radiation. At the time of their formation, the photons would have had a wavelength of about one-thousandth of a millimeter. As the universe expanded, the wavelength of this original radiation stretched along with space. The effect is known as the Doppler

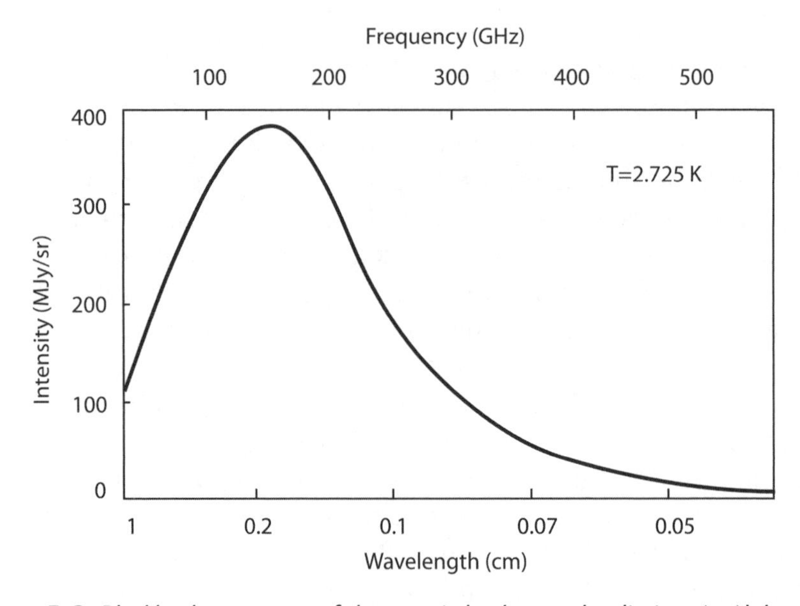

Figure 3-2. Blackbody spectrum of the cosmic background radiation. As Alpher and Herman discovered later, the background radiation occurs in the microwave region of the spectrum. The spectrum of the big bang is known as the cosmic microwave background (CMB) radiation spectrum.

effect and is a fairly common occurrence. The drop in the frequency (which is related to the pitch) of a car's engine after it passes in front of us and moves away, as compared to the higher frequency when it was approaching, is a good example. A large frequency corresponds to a shorter wavelength and vice versa. The following analogy may help explain this. In one minute, you'd count more short train cars than longer ones as a train speeds by you at a crossroad. The shorter cars go by with higher frequency. It's the same with waves. The sound waves you receive from a car receding away from you are longer than if the car is stationary or coming toward you. Therefore, the Doppler effect tells us that as the universe expands and the galaxies move away from each other, the light that we detect is Doppler-shifted, resulting in stretched, longer wavelengths or shorter frequencies. In the case of the universe, the wavelength or the background radiation has stretched from one thousandth of a millimeter to about two millimeters. The current wavelength of the primordial photons is in the microwave region of the electromagnetic spectrum.[16] Alpher and Herman calculated that this background radiation would be of the order of 5 K, a value remarkably similar to the actual value known today. However, they—or anyone else who heard about it—did not think that this background radiation could be identified and actually thought that it had simply mixed with the radiation generated by the stars.

Gamow, Alpher, and Herman, along with James W. Follin Jr., continued developing and improving their work over the next several years. In a paper published in 1953,[17] Alpher, Herman, and Follin presented a fairly complete description of big bang nucleosynthesis of the light elements, which comes very close to present-day understanding.

The 1953 paper is the final publication of the Gamow group. They had achieved early success with their accurate calculation of the hydrogen and helium relative abundances. They also realized that the five-particle island couldn't be crossed and that Gamow's original idea that all the elements had been formed during the first moments of the universe by neutron capture wasn't possible, and that heavier element formation would require a different mechanism. Their discovery of the cosmic background radiation would turn out to be one of their greatest discoveries. But their greatest achievement is perhaps the establishment of the scientific framework of the big bang theory.[18]

Echo of the Big Bang

After their 1953 paper, the Gamow group separated and went on to work on other projects. The difficulties with the five-nucleon island and the apparent impossibility of detecting the cosmic background radiation that they had predicted contributed to the growing disinterest in the big bang theory that ensued. The five-particle island was eventually bridged. As discussed in chapter 1, the heavier elements up to iron are produced in the hot and dense interiors of massive stars. Heavier elements such as selenium, copper, and zinc that exist in our bodies, along with the rest of the heavier elements, all come from supernova explosions. But that wasn't known in 1953 and, consequently, things quieted down in cosmology for a while.

In the 1960s, Robert Dicke at Princeton University led a small group of researchers working on gravitational physics. In 1964 Dicke directed several members of his group to look into the possibility of detecting the background radiation that he thought was present at the birth of the universe. Unaware of Gamow's work, he asked James Peebles to look at the theoretical aspects of the problem, and directed Peter Roll and David Wilkinson to start building an antenna capable of detecting the background radiation. The pair started designing an antenna that would detect radiation of a wavelength of 3.2 cm, which was what Dicke had estimated that radiation to be.

Peebles also went to work, and in about a year, he had figured out the physics of the hot, dense, early universe, independently discovering what the Gamow group had discovered a decade earlier. Peebles's calculations corroborated what Dicke had said, that the universe today was filled with background radiation that originated with the big bang. Like Alpher and Herman had done before, Peebles was able to predict that this radiation should be in the blackbody spectrum.

Thirty miles away, at the Bell Telephone Laboratories communications center in Crawford Hill, New Jersey, Arno Penzias and Robert Wilson had painstakingly calibrated a highly sensitive microwave antenna and were ready to start their measurements of radio sources external to our galaxy. Although they had taken care of removing all sources of electric noise in the electronics, the cables, and the antenna itself, even cleaning up the "white, electrically insulating residue" left over by a pair of pigeons, they couldn't get rid of a persistent signal that seemed to be independent of the orientation of the antenna or of the time when the measurements were taken. The signal corresponded to the radiation emitted by a gas cooled to 3.1 K. Penzias and Wilson continued looking for the source of the noise so that they could get on with their planned work.

In late fall of 1964, Penzias was attending a conference in Montreal and mentioned the annoying problem with the antenna to Bernard Burke of the Massachusetts Institute of Technology. A couple of months later, Burke heard of Dicke and Peebles's work and remembered his conversation with Penzias. He immediately telephoned Penzias and told him to get in contact with the Princeton group.

When Penzias called, Dicke was in the middle of a research meeting with his group. They were discussing the construction of their own microwave antenna with which they were hoping to detect the cosmic background radiation that Dicke and Peebles had predicted should still exist. The next day, the Princeton team was at Crawford Hill meeting with Penzias and Wilson. It became very clear then that the two Bell Labs astronomers had been measuring the cosmic microwave background radiation, the echo of the big bang.

The two teams decided to publish the results of their momentous discovery in two simultaneous papers. The Bell Labs team described the experimental discovery, while the Princeton team explained the theoretical interpretation. The papers appeared in the *Astrophysical Journal* in 1965.[19] Penzias and Wilson won the 1978 Nobel Prize in physics for their discovery.

But the work wasn't done. The agreement between the prediction of what the theoretical values were and the actual values was striking. The Gamow group had calculated that the cosmic microwave background radiation should be about 5 K while the Princeton group came up with about 10 K. These values are considered to be quite close to Penzias and Wilson's measured value of 3.1 K because quantities in cosmology are extremely difficult to measure. However, they had measured the radiation at only one wavelength, at 7.35 cm. And you can't get a spectrum with only one wavelength!

Clearly, values at other wavelengths were needed. They soon came. The Princeton group finished their antenna, which was designed to detect wavelengths of 3.2 cm. Their result fell right on the blackbody curve for a temperature of 3.0 K. The Bell Labs team made another measurement at 21 cm, which had been their chosen wavelength for their original intent of detecting extragalactic radio sources. That was, of course, before their world was turned upside down with their accidental discovery of the echo of the big bang. Their new results at that relatively long wavelength also fell right on the blackbody curve. Other groups joined in the excitement, making measurements at wavelengths ranging from 0.7 mm to 50 cm. All agreed with the blackbody spectrum of about 3 K.

THE COSMIC BACKGROUND EXPLORER

But it wasn't easy digging. The measurements in the millimeter range were extremely challenging to make due to other gas emissions in the atmosphere that were and are also in the microwave region. Experiments were designed to be flown on balloons at high altitudes. These results also agreed with the blackbody spectrum. In the 1980s, a Japanese-American collaboration sent two rockets with detectors calibrated for the very low wavelength range. The first detector sent back results that were consistently higher than the blackbody spectrum values obtained earlier. The second detector malfunctioned due to lack of electrical grounding, which the Japanese engineers were not aware of—a problem that has plagued many space missions and that we're still correcting with recent spacecraft.[20] When some of the data was recovered from this ill-fated mission, it gave results that weren't in agreement with the first mission and agreed with the earlier blackbody measurements.

In November of 1989 NASA launched the Cosmic Background Explorer (COBE) to measure with unprecedented accuracy the cosmic microwave background radiation (Plate VI). COBE was the brainchild of John Mather, a NASA Goddard research scientist who led the team that proposed the design, construction, and launch of four separate instruments on a satellite. The final proposal combined three competing proposals, following a common NASA practice to unite the best ideas from different sources to design the best mission. Along with Mather's team, two other teams were led by scientists who had proposed two of the winning proposals: Luis Alvarez of the University of California at Berkeley, and Samuel Gulkis of NASA's Jet Propulsion Laboratory. The final COBE satellite carried three of the four instruments proposed by Mather: DIRBE, a Diffuse Infrared Background Experiment to search for cosmic infrared background radiation; DMR, a Differential Microwave Radiometer, to map the cosmic radiation sensitivity; and FIRAS, a Far Infrared Absolute Spectrophotometer, to compare the spectrum of the CMB radiation with a precise blackbody.

Most complex missions take many years to design and deploy, and COBE was no exception: COBE took fifteen years from conception to launch. In 1974, when John Mather proposed the mission to NASA headquarters, he was twenty-eight years old, had just received his

PhD in physics from the University of California at Berkeley, and had joined NASA Goddard Spaceflight Center first as a postdoctoral fellow. Shortly after the mission was approved, Mather joined NASA Goddard as a full-fledged research scientist.[21] Initially, the satellite was designed to be launched on a Delta rocket, but NASA decided to switch it over to the space shuttle. In 1986, the satellite was ready for launch when the tragic space shuttle *Challenger* accident took place. The space shuttle program was put on hold pending the results of the presidential commission to investigate the cause of the accident. COBE was again reassigned to be launched on a Delta rocket, and NASA Goddard had to redesign the payload, reducing its mass and volume. Three and a half years later, on November 18, 1989, COBE was launched from Vandenberg Air Force Base in California atop a Delta I rocket on a polar orbit 900 km (560 mi) above the Earth's surface.

The first results from the COBE mission came in soon after. The FIRAS instrument had measured the CMB radiation at different wavelengths, ranging from 0.1 to 10 mm, with a flared horn antenna. The instrument was cooled down to 1.5 K, a temperature lower than that of the radiation it was supposed to measure, to reduce its own thermal emissions. COBE included a reference blackbody that was calibrated with an external blackbody with an emissivity of better than 0.9999. Fig. 3–3 shows the initial set of sixty-seven different wavelengths, spanning the blackbody curve. All the sixty-seven data points fell exactly on the theoretical blackbody curve for 2.725 K, the final CMB radiation temperature that Mather's team calculated, with an uncertainty of one per thousand! The results were astonishing. There was no doubt in anyone's mind that the COBE results had confirmed that the CMB radiation was the smooth, universal blackbody that Gamow, Alpher, and Herman had predicted it should be back in 1948 and that Peebles and Dicke had independently calculated seventeen years later.[22] NASA's COBE satellite had shown us a peek at the formation of the young universe, prompting Stephen Hawking of Cambridge University to proclaim it "the discovery of the century, if not of all time." John Mather won the 2006 Nobel Prize in physics for revealing the blackbody form of the microwave background radiation. George Smoot of the University of California at Berkeley, a member of the COBE team, shared the prize for his discovery of the small fluctuations in the temperature of the radiation.

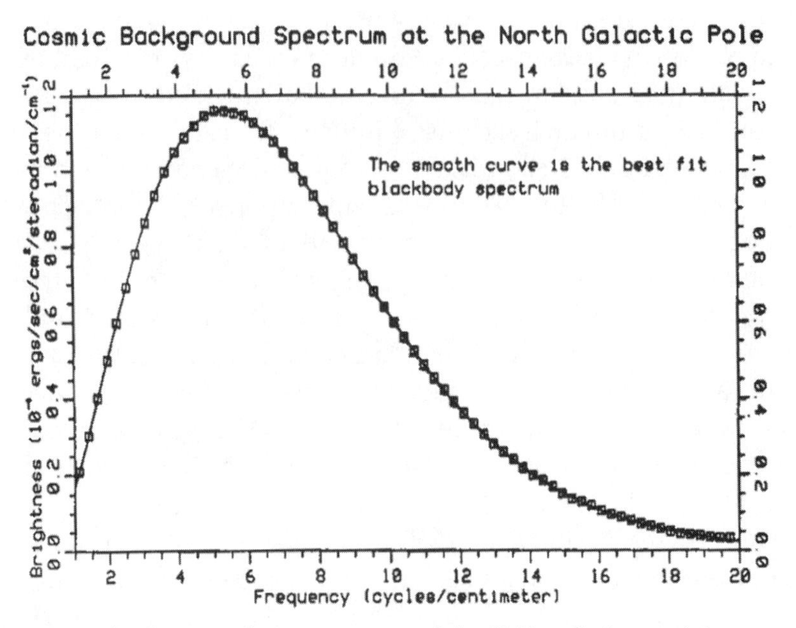

Figure 3-3. The first set of measurements of the CMB radiation at sixty-seven different wavelengths. The data points fall on the theoretical blackbody curve with an accuracy of 1 percent. (Courtesy NASA COBE Science Team)

Lamaître's idea of a universe that originated in an explosion, followed by a rapid expansion that thinned out and cooled the matter, which eventually coalesced into stars and eventually planets, turned out to be not far from the real universe. NASA's COBE satellite had shown us that the primordial radiation that originated during the early stages of the expansion of the universe is still reaching us completely undisturbed. The photons that were collected in 1964 by Penzias and Wilson's horn antenna in New Jersey and in 1989 by the horn antenna on COBE had been traveling along straight lines for 14 billion years. These photons come from the moment, some 300,000 years after the big bang, when the universe became transparent to light and the photons were freed up from the fog of the plasma.

One of the most important predictions of the big bang theory proved to be correct by the COBE results. But that's not the end of the story. As we will soon see, COBE also found that the universe, on a large scale, is very uniform. The almost perfect match of the COBE measurements with the blackbody curve indicates that the big bang must have been extremely smooth and simple. It seems hard to believe that those smooth conditions were produced with an explosion.

Deeper and more difficult problems persist. For Lamaître the big bang was the origin of the universe. There are problems running the movie of the expansion all the way back to time zero. At that point, the big bang theory implies that the universe was infinitely dense and the strength of the gravitational force was infinitely large. But the big bang theory is based on general relativity, which is our modern theory of gravity. And if the gravitational force is infinite, the equations break down. General relativity breaks down at time zero.

Chapter 4

How to Make a Universe

Three Unexplained Problems

The big bang theory is one of the most remarkable achievements of the human mind. With the field equations of general relativity and the present knowledge of the laws of physics, we reconstructed the events that gave rise to our universe, moving temptingly close to the moment of creation. The theory showed us that the universe had an origin, a beginning. Was that the job of a supernatural creator? A remarkable theory called *inflation* has taken us even closer to the first instant and to the answer to this question.

The big bang theory made two very important predictions that were dramatically validated. First, the theory successfully predicted the formation of the light atomic nuclei during the big bang and the correct abundances for these elements. And second, it predicted that there should be a background of microwave radiation filling the entire universe today, the relic of the big bang. The measurements of NASA's COBE satellite spectacularly validated this crucial prediction of the theory.

However, as mentioned in chapter 3, COBE also found that the universe is extremely uniform on a large scale, a difficult thing to imagine coming out of an explosion. The problem is a bit more serious, however. The CMB radiation measured by COBE from all directions was not only similar, as was expected, but indistinguishable to the limit of the equipment, with an accuracy of one part in 100,000!

Recall that the CMB radiation was released some 300,000 years after the big bang, when the universe became transparent to light. Since the universe is about 14 billion years old, two photons arriving now from opposite directions started their journeys toward Earth from two regions in the universe that are now about 28 billion light-years apart. If we calculate what the separation of these two photons was when the universe was 300,000 years old, the time when they were released, we find that this distance is about 90 million light-years. However, at 300,000 years, the field equations of relativity tell us that light could not have traveled more than 900,000 light-years since the big bang. This distance is called the *horizon distance* and it takes into account the speed at which light travels and the expansion of space. Therefore, two points that are separated by 90 million light-years when the universe was 300,000 years old can't be at the same temperature, since no information could have traveled between them. The *light cones* shown in Fig. 4–1 represent the path of these two photons through time. Although the photons were disconnected from each other at 300,000 years, they have reached us today, since they travel at the speed of light. Nevertheless, the uniformity of the CMB implies that these two points were at the same temperature at 300,000 years. This apparent contradiction is called the *horizon problem.*

One possible solution to the horizon problem is to assume that the entire universe began with this extreme uniformity. Although this assumption solves the problem, it does so in an inelegant manner. The fewer ad hoc assumptions that are needed in a theory, the stronger and more valuable the theory is. Aside from that, this assumption makes it extremely difficult to explain the formation of galaxies later on in the evolution of the universe. Galaxy formation requires a nonuniform density distribution that results in a nonuniform gravitational field. A slightly higher concentration of matter in one region of space generates a slightly higher gravitational force toward that region, which in turn pulls in additional matter. The additional mass concentration increases the gravitational field, resulting in additional matter being pulled in. This process couldn't exist with the extreme uniformity of the early universe that COBE has measured.

The horizon problem brings us then to the problem of the *origin of structure.* How did the extremely uniform early universe that COBE observed develop the lumpiness that gave rise to the wide range of structure we see today? Although, on a large scale, the universe is fairly uniform, it is extremely lumpy at shorter scales, in the realm of galaxies and clusters of galaxies (Plate VII). A typical galaxy, like our Milky Way, contains 100 billion stars. Galaxies are grouped in clusters. Our galaxy belongs to the Local Group, a cluster with thirty galaxies. Clusters, in turn, exist in superclusters, and the Local Group belongs to the Virgo Super Cluster. It's likely that many stars are actually solar systems, with their own planets and comets orbiting around them. All of this structure is surrounded by a vast "empty" space—that is, space devoid of stars and other bodies, but still containing minute amounts of matter spewed out in supernova explosions. Even with the enormous number of galaxies in the known universe, the average density of the universe is only about one atom per cubic meter. The problem of the origin of structure is how to explain the formation of this lumpiness if the universe started out extremely uniform.

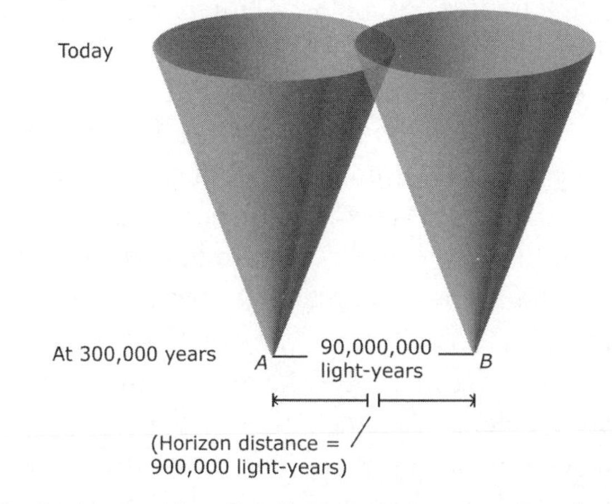

Figure 4-1. The *horizon problem.* Two light cones representing the path of the CMB radiation through time. Two photons received today from opposite directions were released when the universe was 300,000 years old. Because of the expansion of the universe, at that time the horizon distance was 900,000 light-years. Calculations with the big bang equations tell us that these photons were emitted from regions separated by 90,000,000 light-years and were therefore disconnected from each other. There is no mechanism for these two disconnected regions to reach the same temperature. Yet, the photons that we receive today from those two regions indicate that the temperature was the same.

The third problem that the big bang theory does not address is the *flatness problem.* Recall that Friedmann proposed three possible geometries for the universe. Depending on its mass density, the universe can be closed, with a closed or positive curvature like that of the surface of a sphere; open, with negative curvature, like that of a saddle; or perfectly flat, with zero curvature. If the universe is closed, the large mass density will stop and reverse the expansion, turning into a contraction that would end in a big crunch. On the other hand, if the universe is open, it will continue expanding at the same rate forever. If the mass density of the universe is *exactly* equal to the critical value needed for a flat geometry, the gravitational attraction of all the galaxies in the universe will slow down the expansion, but would not be enough to stop it completely. A flat universe will continue expanding forever. Cosmologists have used the letter omega (Ω) as the ratio of the actual density to its critical value, which has been defined as 1. The positive curvature of a closed universe has $\Omega > 1$, while the negative curvature of an open universe has $\Omega < 1$. The zero curvature of a flat universe has the critical value, $\Omega = 1$.

In November of 1978 Robert Dicke gave a lecture on the big bang theory at Cornell University.[1] Interest in his lecture was high since only a month earlier Penzias and Wilson had been awarded the Nobel Prize in physics for their discovery of the CMB radiation.[2] During his lecture, Dicke mentioned the flatness problem. At the time, cosmologists had placed the value of Ω to fall anywhere between 0.2 and 2. At first glance, such a wide range shouldn't have any implications for today's universe. However, Dicke explained that Ω was very sensitive, comparing it to a pencil balanced on its point. Away from any vibrations of air currents, the pencil should remain balanced indefinitely. Any slight tilt in any direction will make the pencil fall. The balanced pencil, said Dicke, represents a universe with Ω equal to 1; that is, one in which the mass density is exactly equal to the critical density. To provide realistic numbers to his audience, Dicke picked the universe at one second of age. That time was important because, according to the big bang equations, that is when nucleosynthesis began. He then showed that for Ω to fall within the range 0.2 to 2 today, the mass density and the critical density at one second after the big bang must have been equal to within one part in 10^{14}. According to Dicke, for the universe that we observe to exist today, the value of Ω must have been no smaller than 0.99999999999999 or the density would have decreased to such a small value that no galaxies could have formed. On the other hand, the value of Ω could not have been larger than 1.00000000000001 or the universe would have collapsed before any galaxies could form.

All three problems—the horizon, the origin of structure, and the flatness problem—are unsolved by the big bang theory. In spite of the tremendous success in predicting the nature of the CMB radiation, magnificently confirmed by the COBE satellite, these unsolved problems mean that the theory is incomplete. As it turned out, even before the COBE satellite was launched, a young scientist would develop a revolutionary new theory that would provide answers to the three unresolved problems and would complement the big bang theory.

Breaking the Symmetry

Sitting among the audience attending Dicke's lecture at Cornell that November of 1978 was a young postdoctoral researcher named Alan Guth. Guth had received his doctorate from the Massachusetts Institute of Technology seven years earlier and had worked as a postdoc ever since, first at Princeton, then at Columbia, and now at Cornell. Guth was determined to pursue an academic career as a professor and researcher in theoretical physics at a major university. The normal route that one takes to achieve this goal is to spend a year or two as a postdoc at a university other than where the PhD was granted, to expand the research interests and to beef up the resume. However, during the decade of the 1970s, jobs for scientists in general and for physicists in particular had become scarce. Young researchers were then stuck in these entry-level, temporary positions for years, waiting for their opportunity to emerge. Such was the case with Guth.

Dicke's lecture made such an impression on Guth that he still remembers many of the details that were discussed. But Guth was a particle physicist and at the moment, he was in the middle of a research project related to the way quarks interact to form protons, neutrons, and other particles. There didn't seem to be any relation to the flatness problem that Dicke had explained in what he was doing.

Two days after the lecture, Guth ran into his friend and fellow postdoc Henry Tye, who told him that the new theories attempting to unify the forces of nature would produce pointlike defects associated with certain esoteric particles as well surfacelike defects that could be very stable. Perhaps, he suggested, they should work on that problem, since these esoteric particles hadn't been observed.[3] Guth didn't know it at the time, but following Tye's suggestion would bring him to solve not only the flatness problem, but the other problems that plagued the big bang theory.

Guth and Tye worked on their new problem and in a span of over a year offered a solution to the overproduction of these anomalies in the early universe. Along the way, they developed new concepts that would soon prove very useful in Guth's development of a new cosmological theory.

Central to the new model is the idea of *spontaneous symmetry breaking*. The concept of symmetry breaking was originally proposed by Werner Heisenberg in 1928 and can be illustrated graphically with a few examples. But first, we need to understand what we mean by *symmetry*. The meaning of symmetry in mathematics and physics isn't too different from what we understand by the word in our everyday language, except that it is more precise. When a system remains unchanged after we perform some operation on it, we say that the system is symmetric under that operation. NASA's space shuttle has symmetry under reflection. If we take a digital photograph and reflect it, the reflected image is difficult to differentiate from the original photograph (Fig. 4–2).[4] The reflected image of a sphere is indistinguishable from the original. A sphere can also be rotated through any angle without changing it. Rotating a cylinder around its longitudinal axis keeps it unchanged. However, a rotation around another axis does change it. This last operation is not symmetric.

Consider now the pencil balanced on its point that Dicke mentioned in his lecture. Like the cylinder of Fig. 4–2, the balanced pencil will not change if we rotate it around its vertical axis (Fig. 4–3); this is a symmetric operation. However, this position is clearly unstable and any small perturbation will topple the pencil, releasing energy. Now, a rotation around the vertical axis will not keep it unchanged. The second operation is not symmetric. Similarly with the marble on the Mexican hat: when the marble is balanced at the center, the configuration is sym-

Figure 4-2. NASA's space shuttle is symmetric under reflection about a vertical axis through its center. A sphere is symmetric under reflection and rotation about any axis through its center. The cylinder is symmetric under rotation through any angle about its longitudinal axis. However, rotating the cylinder about any other axis is not a symmetric operation (space shuttle photograph by the author).

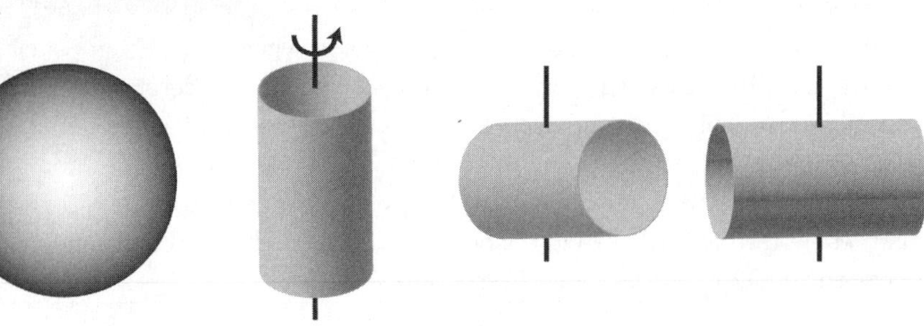

metric under rotation around a vertical axis. This configuration has a higher energy than the more stable configuration when the ball rolls to the bottom of the hat, where the configuration is no longer symmetric.

Heisenberg used the temperature dependence of ferromagnetic materials to illustrate his idea of spontaneous symmetry breaking. A permanent magnet has a north and a south pole that are formed by the alignment of millions of its atoms. The magnet then has a very specific orientation and the force that it exerts on a nearby object made of iron changes if the magnet is rotated. The rotation of the magnet is not symmetric. However, if the magnet is heated above a certain temperature, the alignment of its atoms is lost due to the increased thermal motion and, as a result, the magnet no longer attracts the iron, regardless of its orientation. In this case, rotating the magnet doesn't change the force on the nearby object. Under normal temperatures, the symmetry is hidden or broken, but can be restored with the addition of energy.

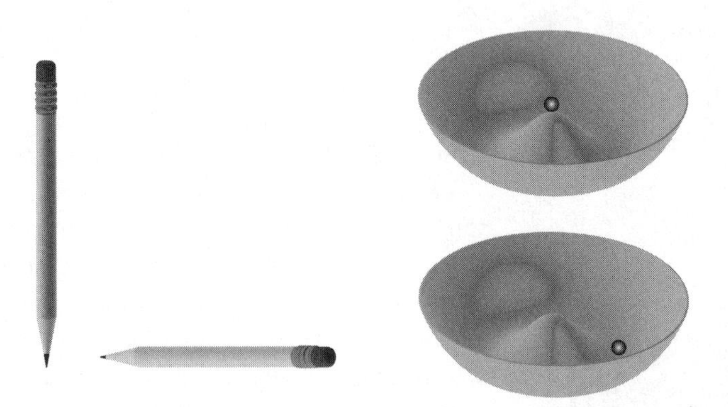

Figure 4-3. Any rotation about the vertical axis of the pencil balanced on its point will not change the configuration. This operation is symmetrical. If the pencil falls, losing energy, the new configuration is no longer symmetric under rotation around a vertical axis. A similar situation occurs with the marble balanced at the center of the Mexican hat at the top right. Any rotation about the vertical axis does not change the system. This configuration has more energy than the configuration at the bottom right, where the ball has rolled to the bottom of the hat. The new configuration is not symmetric under rotations about a vertical axis.

Adding energy to the fallen pencil or to the marble that rolled off the top of the Mexican hat in Fig. 4–3 can restore the symmetry of those two systems. When we add energy to a system in a broken symmetry configuration to bring it back to the symmetric state, the system undergoes a kind of phase transition, just like the phase transition that takes place when ice melts into water as we add energy. Certain systems that are not symmetrical can be described by symmetrical equations when spontaneous symmetry breaking is taken into account. Spontaneous symmetry breaking is carried out with the introduction of a new field called the Higgs field (after Peter Higgs of the University of Edinburgh). The energy of the Higgs field is lowest when the field acquires a nonzero value. Since the state of broken symmetry is lowest in energy (as in the fallen pencil), the Higgs field has a nonzero value; conversely, when the symmetry is unbroken, the energy is highest (the balanced pencil), and the Higgs field has a value of zero.

The Inflationary Universe

In December of 1979, Guth—who was now at the Stanford Linear Accelerator Laboratory (SLAC), having taken yet another postdoctoral position—was finishing up his paper on the solution of the monopole problem with Henry Tye (who was still at Stanford). Remembering Dicke's lecture the previous year, Guth then started thinking about the possible implications of this work to the early universe. Guth assumed that the energy density function of the early universe had a shape similar to the Mexican hat of Fig. 4–4 but with a valley at the center called the *false vacuum*. A ball placed in the hat will most likely settle in the lower circular region, where the energy is minimum. If one were to simulate the effects of temperature by shaking the hat, the ball would jump around. If one were to shake it violently, the ball would move all over the hat and the central peak would become less of an impediment, so that the average position of the ball would be at the center of the hat. If one were to shake with less energy, simulating low temperatures, the ball would settle in the lower circle. However, imagine that the hat was shaken violently, simulating high temperatures and suddenly and very quickly the vibration was slowed down to simulate rapid cooling. Since the average position of the ball at very high temperatures would be the middle of the hat, there would be a good chance that when the vibration was suddenly decreased, the ball would end up at the central indentation. The hat would now be vibrating at a lower temperature, yet the ball would be held up in the central valley, the high temperature region, which is the region of unbroken symmetry or false vacuum.

This situation is actually well known in physics. *Supercooling*, as the phenomenon is known, happens in water when the temperature is lowered at a very fast rate (about a million K per second). At that rapid rate, freezing is bypassed and water remains in the liquid state at temperatures 20 degrees below the freezing point. Supercooled water may actually be the most common form of water in the universe, as it is present in interstellar dust and in the cores of comets.[5] You may have experienced it firsthand, since freezing rain is supercooled water.

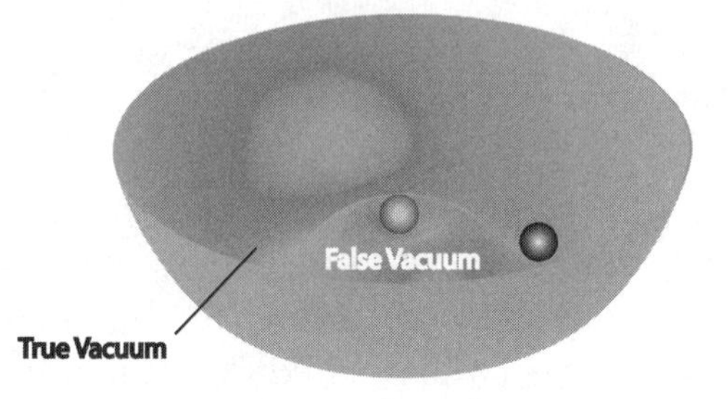

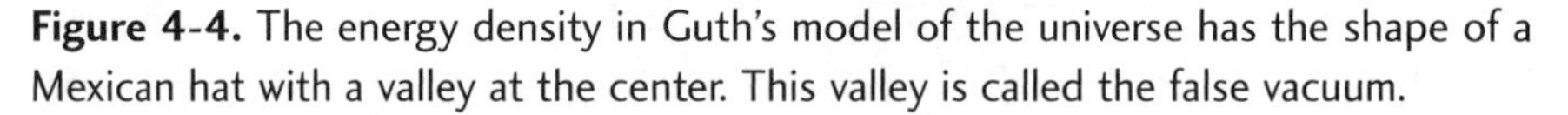
Figure 4-4. The energy density in Guth's model of the universe has the shape of a Mexican hat with a valley at the center. This valley is called the false vacuum.

The early universe had some regions of gas that were at a much higher temperature than the critical temperature of the phase transition, which had been placed theoretically at 10^{27} K. Like the ball that ends up in the unbroken symmetry region at high temperatures, at the higher energies of the early universe, the Higgs field had a value of zero and the symmetry was unbroken. As the universe expanded, the temperature fell, and the universe began the phase transition to the state of broken symmetry, where the Higgs field acquired a value. However, if the cooling rate due to the expansion was much higher than the time required for the phase transition, the universe could have supercooled, reaching temperatures well below the critical temperature with the Higgs field, and remaining at the symmetric zero value in the false vacuum.

It is well known that supercooled water is unstable. For example, when airplanes fly through cirrus clouds, they often encounter droplets of supercooled water that freeze immediately upon contact with the airplane's wings. A de-icing system is then activated to remove the ice buildup on the wings. Would the early universe in the supercooled state of the false vacuum be also unstable? The energy of the false vacuum is higher than the energy of the true vacuum, where the Higgs field has a nonzero value. However, like the ball trapped in the middle valley in Fig. 4-4, the supercooled universe couldn't decay to the true vacuum due to the barrier ring around the false vacuum. At those low temperatures the vibrations wouldn't be strong enough to jump the barrier. In the supercooled water case, it would take some external perturbation to fall from the supercooled state and freeze the water (contact with the airplane wing, for example). The ball in the Mexican hat would need a small external jolt to make it jump over the barrier. If the supercooled state of the universe is unstable, what would make it jump? There clearly isn't anything external to the universe that could do the job!

The answer lies in quantum mechanics. Classically, the supercooled early universe is stable since the lower energy of the true vacuum is separated by a barrier. However, in the quantum mechanical version of the model, the false vacuum is not stable. According to quantum mechanics, the Higgs field is fluctuating and occasionally, in a small region of space, it "tunnels" through the barrier, forming a *bubble* of broken symmetry in the false vacuum. The bubble begins to grow rapidly and takes the false vacuum to the broken symmetry phase, that is, to the true vacuum. Tunneling is a well-known phenomenon that has not only been measured but that also has technological applications that are even commercially available.

The bubble of true vacuum that is formed in the false vacuum grows rapidly because the energy density of the true vacuum is lower than that of the false vacuum. Since the bubble grows, the pressure inside must be greater than the pressure outside, which means that the pressure of the true vacuum that formed inside the bubble is greater than the pressure of the false vacuum, where the bubble initially formed. However, since the pressure of the true vacuum is zero, the pressure of the false vacuum must be negative.

Negative pressure! That's a bizarre concept, as Guth himself has stated.[6] This is not the same as negative *gauge* pressure, which is a relative pressure measured with respect to a standard pressure, usually that of the Earth's atmosphere at sea level. The negative pressure of the false vacuum is measured in relation to the zero pressure of the true vacuum. However, this pressure does not result in mechanical forces, like normal pressures do, since mechanical forces arise only from pressure differences.

In Newtonian physics, gravitational fields are produced only by masses. In relativity, however, mass and energy are equivalent and any form of energy contributes to the total mass. In general relativity, pressure also creates a gravitational field, although its contribution is so small that it becomes negligible under normal terrestrial circumstances.[7] Not so during the early universe, when the pressures are enormously high. The negative pressure of the false vacuum would create repulsive gravity—antigravity!

This reasoning prompted Alan Guth in December of 1979 to perform some calculations to see what this repulsion did to the early universe. After a few hours, he realized that the gravitational repulsion of the false vacuum would cause the universe to expand exponentially, doubling in diameter every 10^{-37} second. That's an extraordinarily rapid rate. Since the supercooled state is unstable, the false vacuum would decay to the true vacuum by tunneling and the rapid expansion would end after 10^{-35} second to be replaced by the standard big bang expansion. In that short interval of time, the universe would have doubled 100 times, increasing in size by a factor of 10^{30}. With these staggering results, I am surprised that Guth didn't simply throw his notebook in the trash and decide that something was terribly wrong with his calculations. Instead, he christened his discovery, calling the rapid expansion of the early universe *inflation*. Then he went home at 1:00 a.m. that night, had a good night of sleep, and couldn't wait to get back to his office the next morning to work out the details of his discovery. That morning, he sat down at his desk and wrote in his notebook:

> SPECTACULAR REALIZATION:
>
> This kind of supercooling can explain why the universe today is so incredibly flat—and therefore resolve the fine-tuning paradox pointed out by Bob Dicke.[8]

Inflation did explain why the geometry of the universe was expected to be flat. Recall that a universe with flat geometry has a value of $\Omega = 1$. Before the inflationary period started, the universe could have had any value, and it likely did. But after one hundred doublings, the value of Ω moved toward 1, changing by a factor of 10^{30} for pretty much the same reason that the surface of a balloon gets flatter as it is inflated (Fig. 4–5). The universe itself, through the inflationary model, provided its own fine-tuning to make Ω equal to 1. There was no need for an external manipulator to fine-tune the universe, which was what was bothering Robert Dicke. Guth felt that this mechanism was so efficient that he was ready to make a prediction: when the appropriate technology to directly measure Ω becomes available, the value that we should obtain would be exactly 1. We should find that the universe today is completely flat.

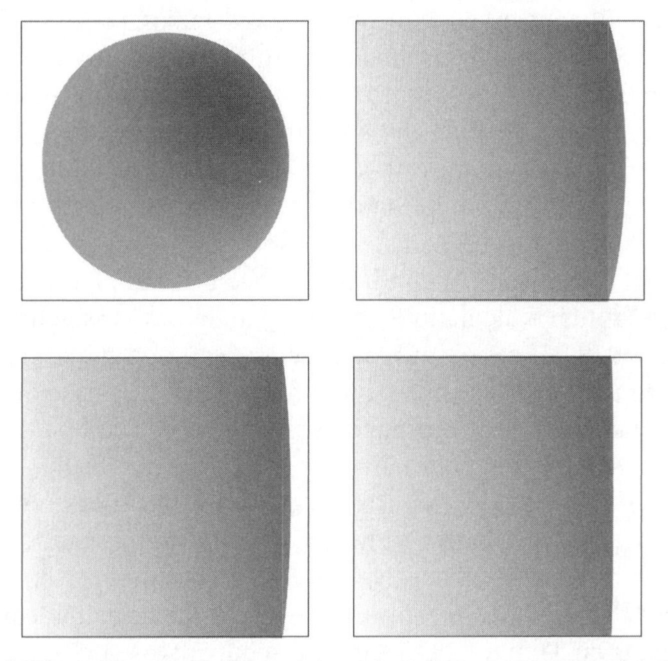

Figure 4-5. These four panels illustrate how the inflationary model of the early universe solved the flatness problem. The sphere in the first panel is inflated by a factor of about 3 in each subsequent panel. By the fourth panel, the section of the sphere that fits in the panel has a curvature that is indistinguishable from a flat geometry. In the short period during which inflation took place, the universe doubled in size one hundred times.

Guth's inflationary model also solved the horizon problem, since the entire observed universe of today was a tiny fraction of the horizon distance during the early universe. Therefore, the whole region could reach thermal equilibrium, becoming homogenized. At the end of the inflationary period, the homogenized region reached a diameter of 10^{400} light-years, while the observable universe then was only about 1,000 cm. The two regions continued expanding with the big bang and today, the observable universe is a very small fraction of the homogenized horizon distance. The inflationary model explains the extreme uniformity of the CMB radiation: the primordial radiation that we receive today from all parts of the observable universe was emitted from a homogenized region in the early universe.

Spectacular realization, indeed! With his inflationary model of the universe, Guth explained two of the three major problems encountered by the big bang. What about the third? If the inflationary mechanism stretches the universe to that degree, producing homogenous regions that are larger than the known universe, how does it generate the small inhomogeneities that are needed for the formation of galaxies? It seems as if inflation didn't solve the problem of structure. More work needed to be done.

But that could wait. After all, Guth came up with all the main ideas and some rough calculations for his model in a single night in December 1979. Of course, all the previous work on the esoteric particles, which included the concept of the supercooling phase of the early universe, had been done during the previous year. The first thing to do now was to probe his colleagues for their reaction to his discovery. Could all of this be nonsense? Had he overlooked some obvious and simple thing that would not allow his mechanism to work?

His colleagues reacted positively to his succinct description of his discovery. He then decided to break the news to the scientific community in a seminar on January 23, 1980. The seminar was successful beyond all his expectations. The news of his revolutionary discovery spread relatively quickly throughout the physics and astronomy community at the major universities. That brought Guth a more immediate problem to solve. It seemed as if his postdoc days would be ending, as offers from many of the major universities began to come in. Eventually he settled on his alma mater, MIT, where he is now a professor of physics. His paper on the inflationary model of the universe appeared in 1981.[9]

THE NEW INFLATIONARY UNIVERSE

At MIT, Guth continued working on his model, trying to understand the process of bubble formation during the inflationary period and the possibility that these bubbles could provide the necessary inhomogeneities needed for galaxy formation without compromising the precious uniformity that the model explained so well. The difficulty was that the bubbles formed during the phase transition would not all coalesce, remaining in separate clusters each dominated by a giant bubble where most of the energy would concentrate. The inhomogeneity that resulted was not what was observed. By this time, however, Guth wasn't alone anymore. His model had caught the attention of several of the top cosmologists, and several groups were now attempting to come up with solutions to the inflationary theory's problems.

As it turned out, Guth hadn't been alone ever. Andrei Linde of the P. N. Lebedev Physical Institute in Moscow (now at Stanford University) had independently developed very similar ideas to those of Guth, and at about the same time. However, Linde did not see that the exponential expansion would solve any of the problems with the big bang and decided that "there was no reason to publish such garbage."[10] However, after Guth's paper showed the viability of the model, Linde returned to it, successfully developing a new approach that preserved the basic concept of Guth's model and at the same time solved the problem of structure. And this time, he published.[11] Independently of Linde's work, Andreas Albrecht and Paul Steinhardt of the University of Pennsylvania developed a very similar new model.[12]

The new inflationary theory uses a slightly different form of the energy density function describing the Higgs field (Fig. 4-6). The new function has no energy barrier and the false vacuum sits on a flat plateau. Instead of quantum tunneling, the Higgs field slowly rolls down from the false vacuum to the broken symmetry of the true vacuum. If the plateau were flat, how would the field begin to roll? It turns out that thermal or quantum fluctuations are present in this region and it is those fluctuations that can get the field rolling. The process is similar to that of a ball sitting in equilibrium on a flattened Mexican hat with the same shape as this energy density diagram. If we gently shake the hat to simulate thermal fluctuations, the ball will slowly begin to roll down the hill, reaching the lower part of the hat's brim, where it oscillates back and forth until friction slows it down and it finally stops. As in the original inflationary model, a bubble of broken symmetry forms in the false vacuum. In the new model, however, the bubble starts to grow by many orders of magnitude while it is still moving along the plateau, where the energy density is high. By the time the bubble slides to the trough of the true vacuum, it has enlarged to such magnitude that it can enclose the entire observed universe. This rapid expansion has another important effect: it smoothes out any inhomogeneities that may be present within the bubble. And since the bubble encloses all of the observable universe, this mechanism ensures that the universe starts out smooth and uniform.

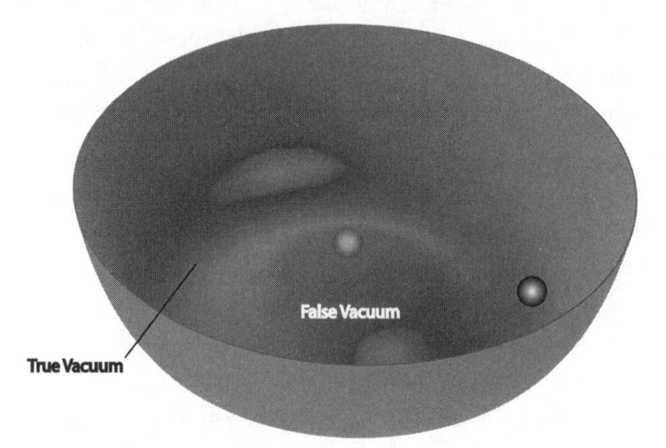

Figure 4-6. The energy density in the new inflationary model of the universe has the shape of a flattened Mexican hat. The plateau is the false vacuum. Thermal or quantum fluctuations push the Higgs field from its symmetric state in the false vacuum to its broken symmetric state of the true vacuum at the trough.

But wait! Wasn't that one of the problems that the original inflationary model didn't solve? If the universe starts out perfectly smooth and homogeneous, how do we get the lumpiness that is needed for the formation of all the structure that we see in our universe today—the stars, galaxies, clusters, and superclusters? It appeared as if the new inflationary model was too successful at erasing the nonuniformities that would end up in galaxies.

The solution to this problem was the result of the brainstorming by a group of scientists gathered for three weeks at the Nuffield Workshop on the Very Early Universe held in Cambridge, England, in 1982. After many discussions among the group of scientists that included Stephen Hawking of Cambridge University, Michael Turner of the University of Chicago, A. Starobinsky of the Landau Institute of Theoretical Physics in Moscow, as well as Guth, Linde, and Steinhardt, the solution became clear. The lumpiness is originated by quantum fluctuations of the Higgs field during the phase transition. According to quantum theory, all physical quantities fluctuate. The fluctuations of the Higgs field take place in regions that are small enough for quantum phenomena to be important. The inflationary process enlarges them to the macroscopic scale, where they can act as seeds for the formation of structure.

The new inflationary theory solves the problem of structure that the original inflationary theory couldn't solve. New inflation also solves the flatness and the horizon problems in the same way that the old theory did. Because of the exponential rate of inflation, the evolution of the universe appears to be independent of the initial conditions. Before inflation started, the value of Ω could have been almost anything, and, after the early universe doubled one hundred times, Ω was driven toward 1. Any gross inhomogeneities present at the start of the inflationary period were also erased during the extraordinary stretching of space that took place. The new inflationary model assures that the universe starts out perfectly flat, uniform throughout but with the small lumpiness that can serve as the future seeds for structure, so that when this marvelous process ends and the standard big bang processes of nucleosynthesis, expansion, and galaxy formation take over, the universe ends up correctly.

The Fingerprint of the Early Universe

At the Nuffield Workshop, Hawking, Guth, Steinhardt, Turner, and others derived the spectrum of the density perturbations using the general relativity equations for the inflationary model. The spectrum of perturbations generated by quantum fluctuations was of crucial importance for the validity of the new inflationary model. If the quantum fluctuations generated an energy distribution that was too lumpy or not lumpy enough, they would not end up in galaxies. The fluctuations had to be just right for the right universe to evolve. The problem was not an easy one to solve and the scientists worked on it independently, periodically checking their results with one another. Hawking, true to form, worked on his own and would only report his results during one of his talks, giving only a sketchy idea of the method that he'd used and without explaining why his result had changed. The others were more open in their communications. Steinhardt and Turner decided to use a more rigorous method developed by James Bardeen of the University of Washington. Guth decided to use an approximate method but was the only one getting consistent answers throughout. Finally, on the last day of the workshop, their answers converged to the same result (the one that Guth kept getting). The spectrum obtained was what is called a *scale-invariant* spectrum. This type of spectrum appears to be a random pattern. However, if the pattern is invariant, it can be decomposed into a sum of sinusoidal waves with the same heights (Fig. 4–7). According to the inflationary model, the waves with shorter wavelengths would be produced toward the end of inflation because the decreasing energy at this end would create smaller quantum fluctuations. The spectrum would reflect this phenomenon, and the heights of the waves would get smaller as the wavelength would decrease. Scientists call this phenomenon *tilt*.

With their calculation at Nuffield, the cosmologists developed a fingerprint of the radiation pattern of the early universe, a pattern that would have the form required for the seeds of galaxy formation. At the time, none of the workshop attendants thought it was feasible to obtain real data to corroborate this very specific prediction. It never crossed their minds that they were placing the new inflationary theory on the line for a hard test. They were in for a surprise.

At about the time of the workshop, NASA gave a green light for the development of the COBE mission, which launched in November 1989. This mission would actually collect precisely the data for the spectrum that the Nuffield cosmologists had predicted. Recall that within a couple of months of the satellite deployment, NASA had reported the results of its measurement of the microwave background radiation at different wavelengths. This radiation had been emitted when the universe became transparent to light, 300,000 years after the big bang. The spectrum reproduced perfectly the blackbody radiation curve of 2.725 K that the big bang theory had predicted.

A perfectly smooth blackbody radiation curve would not have allowed for the formation of tiny concentrations of matter during the early universe that would seed the universe. But COBE also carried a Differential Microwave Radiometer (DMR), which would enable it to map the sensitivity of the cosmic radiation and to measure the small variations in the CMB radiation.

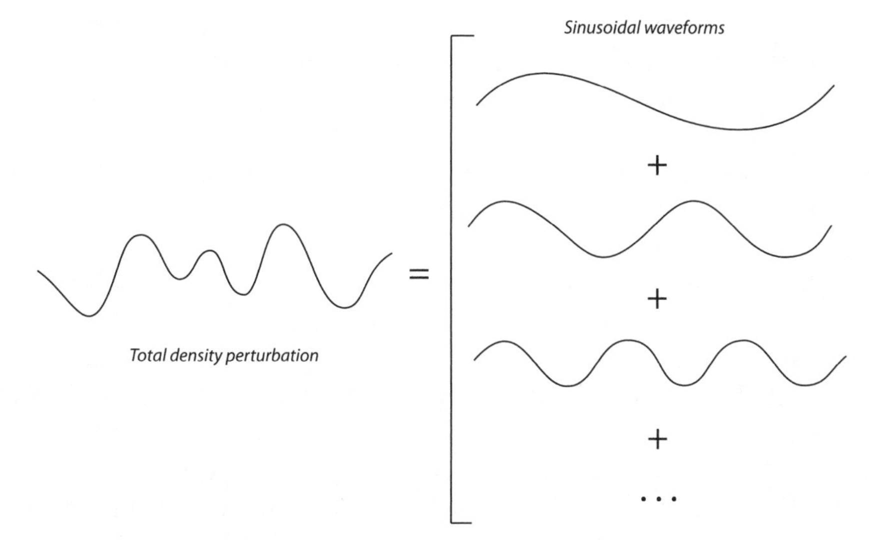

Figure 4-7. The spectrum of density perturbation is a scale-invariant spectrum, which can be decomposed into a sum of sinusoidal waves all with the same height. The inflationary theory predicts that the spectrum also has tilt. This scale-invariant spectrum with tilt is the fingerprint of the early universe.

By April 1990 the DMR had completed its first survey of the entire sky. The first two analyses showed no variations at the level of 1 in 10,000. They found no signs of any seeds that could have eventually formed galaxies. The instruments were so precise and the measurements so delicate that engineers had to account for the effect of the Earth's magnetic field on the instruments on board. The possible currents that the Earth's magnetic field can generate on metals are normally negligible, but in this case, they were of the same order of magnitude as the effects that the instruments were attempting to detect. And even with that precision, there were no signs of any wrinkles or variations in the density of the primordial universe that could have made stars and galaxies. "We haven't ruled out our own existence yet," said John Mather, the NASA principal investigator. "But I'm completely mystified as to how the present day structure exists without having left some signature on the background radiation."[13]

Months went by and still no signs of any variations in the CMB radiation were reported. The scientists outside the COBE team were beginning to worry about the lack of results, especially those who were working on the inflationary theory that had made a definite prediction not only on the need for the variations, but on the actual shape of the spectrum to be obtained. As it turned out, more precise data had been coming in but since it was still tentative, NASA had decided to hold its release until its scientists were completely sure that the data were real. In the end, NASA's insistence on reviewing and checking the data with extreme caution paid off, and the team was satisfied that they had done it right. George Smoot, the principal investigator for the DMR, was so sure of the correctness of the data that he offered a plane ticket to anyone who could find a mistake in their interpretation of the data.

NASA made the announcement at the American Physical Society annual meeting in Washington in April 1992. Many of the scientists working on the inflationary theory were present, eager to hear the results. Smoot distributed the COBE spectrum to the audience right before beginning his presentation. The COBE satellite had detected cosmic microwave background fluctuations of one part in 100,000 (Plate VIII). These tiny variations were believed to have been the seeds for the formation of the stars, galaxies, clusters of galaxies, and all the structure that we observe today. But what was more important to the inflation theorists, the DMR spectrum showed a complete agreement with the shape of the predicted scale invariant spectrum of the density perturbations that they had calculated during the 1982 Nuffield workshop.

"FROM SPECULATION TO PRECISION SCIENCE"

The COBE mission was an enormous success, providing hard evidence for the predictions of the big bang theory through its precision measurement of the microwave background radiation and for the inflationary theory with its discovery of tiny fluctuations (or anisotropy) in the CMB. But NASA didn't sit on its laurels. In June 2001, it launched the Wilkinson Microwave Anisotropy Probe (WMAP) from Cape Canaveral to map the temperature fluctuations of the CMB radiation with much higher precision and sensitivity than COBE had.

The first results from WMAP were announced in February 2003, nineteen months after its launch. By now, the confirmation of the predicted density fluctuations from the COBE mission had convinced most cosmologists that the inflationary model of the universe was probably correct.[14] The inflationary theory had made three important predictions. The first prediction had already been confirmed by COBE: the CMB radiation should have a spectrum of density perturbations that was scale invariant. The second prediction was that Ω should be exactly 1, which implies that space is perfectly flat. The third prediction was the detection of the tilt in the spectrum of density perturbations.

Scientists working on cosmology, students, and people interested in astronomy from all over the world gathered in packed auditoriums to watch the press conference being broadcast from NASA Headquarters in Washington. Mission Director Charles Bennett showed the first image: a large map of the entire sky compiled from a yearlong exposure (Plate IX shows the same image with three years of data that were released in 2006). This image was a detailed photograph of the infant universe!

The photons that formed the image in WMAP's instruments started their journey 13.7 billion years ago. They came from the primordial epoch, when the universe became transparent to light, 300,000 years after the big bang, traveling along straight lines until they reached WMAP's detectors. And now thousands of people from all over the world were marveling at this technological achievement.

The marvelous WMAP image showed temperature variations of the CMB radiation with much higher resolution than COBE. The pattern of density fluctuations fits the scale-invariant spectrum developed by the inflationary-theory cosmologists at Nuffield. The new mission corroborated, with greater precision, what COBE had found. The first prediction of the inflationary theory was still safe and sound.

To test the second prediction about the geometry of space, the WMAP mission scientists selected two regions on the image and, since they knew the distance that the photons had been traveling since their birth, they drew a big triangle in the sky with sides joining the two regions and the Earth. Then they measured the angles with high precision. The result? Exactly 180°. Space is flat, as the inflationary theory predicts.

If space is flat, the density of the universe is exactly equal to the critical density. The value of Ω is precisely 1. But the universe didn't have to start with that value. That would have been difficult to accept, because it would have smelled of fine-tuning. That's what bothered Robert Dicke and would have bothered the scientific community. But the beauty of the inflationary theory is that ending up with a value of exactly 1 doesn't require any fine-tuning. Any value works just fine. The period of inflation drives that value toward 1 and hands it to the big bang, to expand the universe with a perfectly flat space.

The analysis of the data for the third prediction, the tilt in the spectrum of the CMB radiation, would need more time.

The science team went back to work. But they had achieved more than enough. "This," said John Bahcall of the Institute for Advanced Study in Princeton, "is a rite of passage for cosmology, from speculation to precision science."[15]

In March of 2006, they were ready with their second announcement. To see if there was any tilt in the pattern, the WMAP science team had compared the brightness of short-distance fluctuations given by the broad features to the long-distance waves given by the compact features in the afterglow light. They saw that the relative brightness decreased as the features got smaller, which was in total agreement with the prediction. The theoretical fingerprint developed at Nuffield over many hours from the general relativity equations turned out to be the fingerprint of the real universe. It's hard to imagine a more stringent test for a theory, and the new inflationary theory passed it with flying colors.

MAKING THE UNIVERSE

The COBE mission verified the prediction obtained with the general relativity equations for the big bang that the CMB radiation has the smooth, universal blackbody curve. COBE and later WMAP verified all the predictions of the new inflationary universe theory. Which theory is the right one? Both are. Inflation acted only during the first 10^{-35} second of the life of the universe. During the unimaginable short time of a billionth of a billionth of a trillionth of a second, it increased the size of the universe by a factor of 10^{30}. Very shortly after its origin, as the temperature of the universe began to come down, the universe settled in an unstable, supercooled, and symmetric state called the false vacuum. This state behaves as if it has negative pressure and, according to general relativity, negative pressure produces negative gravity. This negative, repulsive gravity accelerated the rate of expansion until the energy and mass of the universe changed back to the state with attractive gravity.

When the era of inflation ended, the universe had a perfectly flat geometry, and was homogenous but with very tiny variations in temperature due to quantum fluctuations. The energy that had generated the inflation produced a hot and dense gas. At this point, the standard big bang expansion took over, and the production of the light elements by nucleosynthesis began. A short time later, when nucleosynthesis ended, the temperature was about one million degrees—too high for the nuclei of these recently formed elements to pair up with electrons. The universe was filled with a hot plasma that continued to expand and cool for the next 300,000 years. When the temperature was cool enough for nuclei and electrons to form atoms, the universe became transparent to light. The primordial photons that were liberated at this time traveled undisturbed along straight lines throughout space as the universe expanded. Today, there are about 400 million of the original photons for every cubic meter of space. They form the CMB radiation that entered Penzias and Wilson's horn antenna in New Jersey and that COBE and WMAP detected from orbit.

Work continues to improve these two successful theories to advance our understanding of the way the universe evolved from almost nothing to what we observe today. Scientists are attempting to obtain answers to the many unanswered questions that remain. But the biggest unanswered question is, how did it all begin? We think we have a good idea of how it inflated from an almost undetectable size and how it later on expanded and generated galaxies and stars and planets. But we don't know how this infinitesimal bundle of energy appeared. Was it there all along? If so, how did it remain as such and what made it suddenly inflate to produce our universe? If it wasn't there, how did it appear? The inflationary universe theory does not attempt to explain how the universe emerged in an inflated state because it is a theory of the *evolution* of the universe from an instant after creation to the moment, 10^{-30} second later, when it handed it over to the big bang.

The inflationary theory assumes that the laws of physics already exist and operate in the same way that we know them to operate today. It has no alternative. The laws of physics make the universe possible. They are the watchmaker.

Chapter 5

THE WATCHMAKER

THE LAWS OF PHYSICS

The inflationary theory and the big bang theory are among the most important scientific achievements of the modern scientific era. For the first time in history we are able to peek at the first moments in the life of the universe and understand how it evolved to form galaxies, clusters of galaxies, stars, and the wide variety of structure that constitutes our universe today.

According to the inflationary theory, the universe underwent an extremely rapid acceleration that ended when the universe was 10^{-35} second. This expansion was guided by the fundamental laws of physics, which are assumed to have been already at work, behaving in exactly the same way that we know today. This assumption is nothing more than Einstein's principle of relativity, which says that *the laws of physics are the same everywhere in the universe.* The principle of relativity is at the heart of science since, without it, we simply couldn't discover the way nature works. If the laws of physics change in time or space, the universe would be completely unpredictable. If the laws of physics change, the electromagnetic radiation arriving at our radio telescopes from Jupiter would not tell us anything about the planet's composition or temperature. It is because of the principle of relativity that we are able to analyze this radiation and discover that Jupiter has a rocky core with a mass of about 318 Earths, and that this core is surrounded by a 40,000-km (25,000-mi) layer of liquid hydrogen and an outer layer of gaseous hydrogen 10,000 km (6,200 mi) thick. When one of our spacecraft arrives at Jupiter with an array of instrumentation, we find that these remote measurements were correct and that the laws that we used to make those determinations did not change in time or in space.

The laws of physics make the universe work. *They* are the watchmaker of the universe. But, what are the laws of physics? The laws of physics are the fundamental generalizations obtained from our observation of nature; they describe the underlying structure of the universe. For all its complexity, physics can be reduced to the study of four fundamental interactions. These interactions, or forces, are the electromagnetic, gravitational, weak nuclear, and strong nuclear forces, and they account for everything that we observe in the universe. They describe all the variety of phenomena in the universe: from the explosion of a supernova to the falling of a leaf on Earth, from the motion of a planet around a distant star to the rapid drumming of a woodpecker, or from the collision of two hydrogen nuclei during the nucleosynthesis phase of the early universe to the crack of a wooden bat hitting a baseball. Everything that happens anywhere and at any time in the universe is ultimately controlled by one of these four interactions.

The *electromagnetic force* binds electrons to nuclei to form atoms and binds atoms together to form molecules, holding them together in solids. The electromagnetic force is described by the theory of electromagnetism developed in the nineteenth century by James Clerk Maxwell, Michael Faraday, and a few other scientists. It combines the electric and magnetic forces into one single interaction, since the two are intimately related.

The *gravitational force,* the first interaction to be discovered, was first formulated by Newton with his universal law of gravitation. Gravity controls the fall of an apple, the motion of the Earth around the sun, and the rotation of the Milky Way—and plays an important role in the rate of expansion of the universe. Einstein's general theory of relativity is an extension to Newton's theory of gravity.

The *weak nuclear force* is responsible for the beta decay process. The beta decay of a neutron into a proton and an electron was the first step in the nucleosynthesis era of the early universe. The weak force controls many of the reactions that take place in the sun and the stars.

The *strong nuclear force* holds nucleons together in the nucleus. Like the weak force, the strong force is a short-range force, acting only at distances shorter than 10^{-15} meter. The strong force only acts on protons and neutrons. This force is actually a residual force that reflects the actual interaction between the quarks that make up the protons and neutrons. The strong force is 137 times stronger than the electromagnetic force and a 100,000 times stronger than the weak nuclear force.

In reality, there are other forces in nature, but they don't play a direct role in the way particles make up the universe. One of these additional interactions is the Higgs field that made possible the inflation of the universe and that is involved in giving mass to the fundamental particles. Another is the cosmological constant, which can be thought of as negative gravity and makes the universe accelerate at a faster rate.

THE QUANTUM RULES

All interactions between subatomic particles are governed by the rules of quantum mechanics. These rules are part of the laws of physics. According to quantum mechanics, the world is grainy, quantized; it changes by steps and not continuously. The step size of quanta is controlled by *Planck's constant*. The magnitude of this constant, which has been measured experimentally with great precision, determines that the step size of quanta is very small.

Quantum mechanics tells us that particles in the realm of the atom behave in peculiar ways. If you perform two sequential operations to a subatomic particle and then reverse the order, you'll find that there is a difference in the final result. You'll also find that this difference is always related to Planck's constant. Although this phenomenon might seem strange at first sight, it's really not that uncommon in the world of our experiences. There are certain operations that you can perform to macroscopic objects in which the order matters. For example, if you take a book and lay it flat on a table such that you can read the title and then rotate it toward you, so that now the cover faces you, then rotate it to your right; in the final configuration, you are able to read the title on the book's spine (Fig. 5-1). If, on the other hand, you switch the order of the operations, rotating it first to the right and then toward you, the book will end up with the cover facing you but with the spine against the table. In this case, the order in which you performed the rotations gave you a different result. It's the same with subatomic particles.

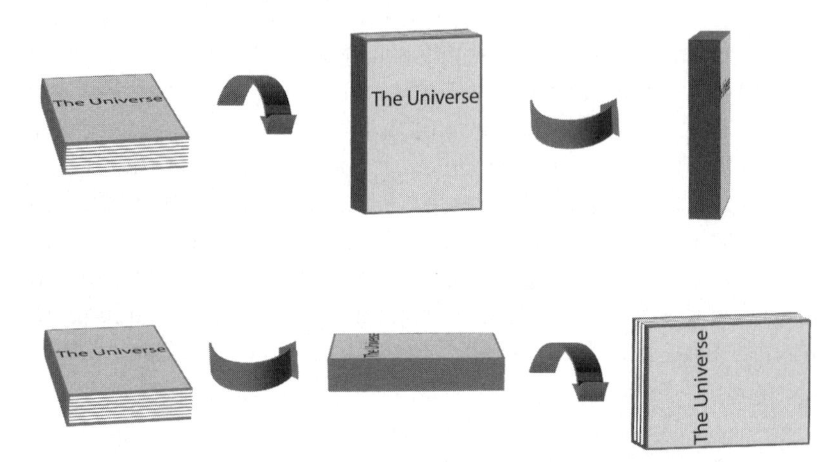

Figure 5-1. The order in which the two rotations of a book are performed changes the final orientation of the book. According to quantum mechanics, the order in which certain operations are performed on a subatomic particle also matters. The final outcome is different for the two sets of operations and the difference is always proportional to Planck's constant.

Light also has a peculiar behavior. According to quantum mechanics, the nature of light (and of all electromagnetic radiation) is such that it reveals itself as having the properties of a wave or those of a particle, depending on the type of experiment that we perform. When light manifests itself as a wave, it has most of the familiar properties of mechanical waves, such as wavelength, frequency, and amplitude, although it travels at a constant speed regardless of the motion of the observer. When the experiment that we perform shows light as a particle, a photon, light has the properties of particles.

In 1982, Alain Aspect of the Centre National de la Recherche Scientifique in Paris performed such an experiment. He triggered a calcium atom to emit a single photon by first energizing it to a specific energy level. In this excited state, the electron sits on the higher-energy level as if it were a marble that has been kicked up to the edge of a step in a staircase. At the very edge of the step, the marble is unstable and at any moment can fall back down to the bottom of the staircase. In the case of the atom, the electron falls back to its original energy level in just 4.7 nanoseconds, releasing a photon with energy equal to the difference in energy between these two levels.

In his experiment, Aspect sent the single photons he obtained toward a beam splitter, a research-grade two-way mirror that transmits half of the light that strikes it and reflects the other half. The beam splitter was positioned at a 45-degree angle with the direction of the incoming photon (Fig. 5–2). He then placed detectors in two possible paths of the photon, the transmitted and the reflected. With this configuration, a wave would split into a transmitted component and a reflected component. A particle, however, would either be transmitted or reflected. It couldn't do both at the same time. In Aspect's experiment, with the detectors in place, the photons acted like particles. They were either transmitted or reflected.

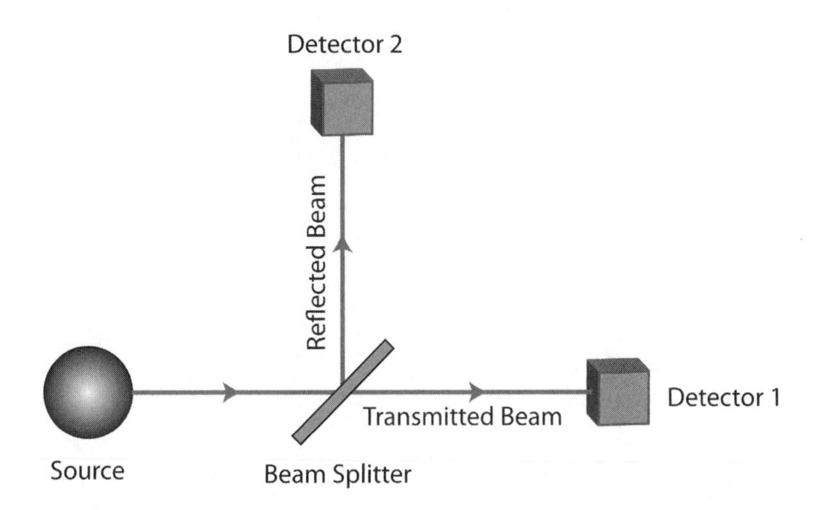

Figure 5-2. Alain Aspect's experiment to determine whether single photons would be reflected or refracted by a beam splitter. By placing detectors in the two possible paths, he was able to measure the particle properties of the photons.

Aspect replaced the detectors with regular mirrors to combine the light from both paths (Fig. 5–3). Without the detectors behind the beam splitter, he couldn't tell which way the photons went. He only saw that photons arrived at either one of the two screens. After many photons had gone through, one at a time, he began to see an interference pattern form on both screens, the result of the reflected wave of light combining with the light bouncing off the full mirror or the transmitted wave combining with the light from the second full mirror. The photons that were emitted from the calcium atom were now producing interference patterns, the regions of reinforcement and cancellation formed when two waves met. Interference patterns are the hallmark of waves.

Elementary particles also reveal themselves as having the properties of a particle or those of a wave. In 1923, a French graduate student by the name of Louis de Broglie extended the dual properties of light to electrons and calculated their wavelength. A few years later, electron waves were detected experimentally. Electron waves are the basis for the scanning electron microscope. As with light, the particle nature of the electron and other elementary particles reveals itself in experiments attempting to detect particles, while the wave nature appears when the experiment is designed to detect waves.

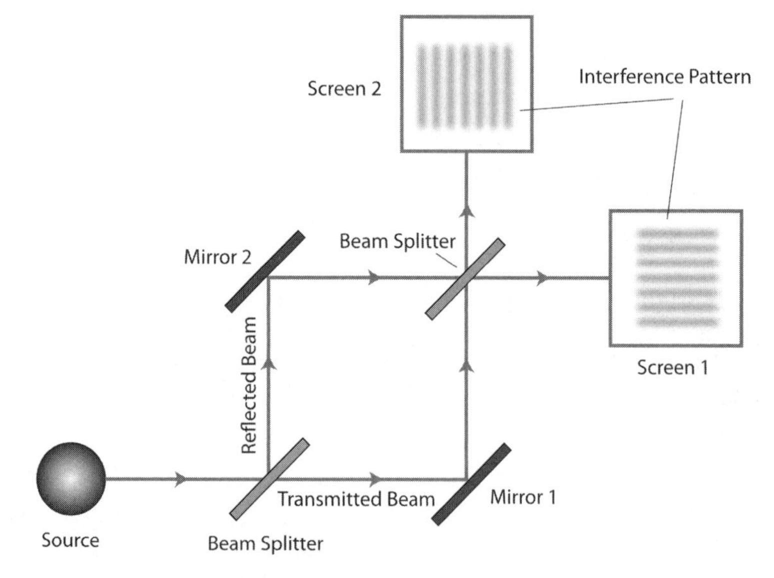

Figure 5-3. In a subsequent experiment to the one shown in Fig. 5-2, Aspect replaced the detectors with regular mirrors. Without the detectors in place, he couldn't determine the path of the photons. The photons formed an interference pattern on the screens, exhibiting their wave behavior.

To see how this phenomenon works with particles, consider an electron gun, like the ones used in the old-style CRTs or television sets. If you aim the gun at a screen where two narrow slits have been cut out, the electron beam splits into two narrower beams that strike a screen placed behind, forming an interference pattern (Fig. 5–4 Left).

The setup was configured for the two beams to interfere with each other like waves and we detected waves.

Imagine now that we attempt to observe the electrons as they cross each slit by placing small detectors behind the first screen (Fig. 5-4 Right). The detectors send a weak electromagnetic signal that bounces back toward it when an electron is detected. When we start the electron gun again, the two detectors register a signal as single electrons cross one or the other slit and go on to hit the screen. After a while, the pattern of collisions on the screen is just what is expected of particles: most strikes appear in front of each slit. But no interference pattern is formed. The detectors were placed so that we can observe the electrons as particles, and we detected particles, not waves. Without detectors in place, we see waves; with the detectors in place, we see particles.

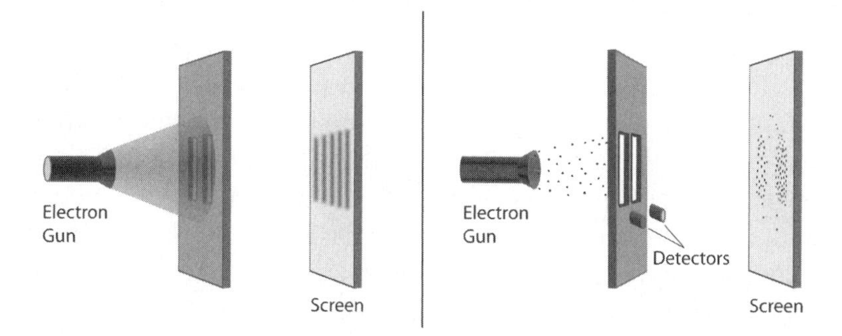

Figure 5-4 (Left). Electrons from an electron gun pass through two narrow slits and strike a screen behind. **(Right).** Small detectors send a signal every time an electron goes by the slit. In this case, the electrons behave like particles.

Suppose now that we attempt to fool the system by reducing the intensity of the signal that the detectors send, since we suspect that our signal is disturbing the electrons. If we reduce the intensity of the light until we have only one single photon for every electron that passes through the slit, we can account for every electron and no interference pattern appears. If we reduce the intensity even more, there will be fewer photons than electrons passing and some electrons will be unobserved. When we do that, a faint pattern begins to appear. The unseen electrons behave as waves and form the pattern, and the detected ones move on along straight lines and strike the screen.

Reducing the intensity reduces the energy of the photons. According to quantum mechanics, the energy of each photon is proportional to the frequency of the light. So a decrease in intensity results in a decrease in frequency, which implies an increase in wavelength.

Now, an increase in wavelength reduces the resolution in our detector; that is, the ability to locate the electron (this is similar to attempting to pick up grains of salt with our hands instead of using tweezers). When the intensity is too low, the wavelength becomes larger than the distance between the slits and we aren't able to see which one of the slits was the one the electron went through. In this case, the interference pattern returns.

Our inability to observe the electrons without disturbing them didn't work. What quantum mechanics teaches us is that this scheme *cannot* work. Our failure isn't due to lack of more precise instrumentation or better techniques; it's an intrinsic limitation of nature. When we use a short enough wavelength to locate the position of the electrons that pass through the slit, there is an increase in the energy of the photon that interacts with the passing electrons. A more energetic photon produces a larger uncontrollable recoil velocity on the electron it interacts with. When we try to reduce the recoil velocity of the electron by reducing the energy of the photon, the wavelength increases and we lose information on the location of the electron. There is connection between our attempt to measure the position and the velocity of an electron.

In 1927, the German physicist Werner Heisenberg formalized this intrinsic limitation of nature in a paper titled "On the Perpetual Content of Quantum Theoretical Kinematics and Mechanics."[1] This mathematical formalization became known as the *Heisenberg uncertainty principle*. In an article that he wrote for a German magazine two weeks after the submission of his paper, Heisenberg explained that electrons could not be described using common language or concepts, such as position or velocity.[2] Even though we call electrons "particles," an electron is nothing like a dust particle. Electrons and other elementary particles behave like particles and like waves, depending on the specific observation we make. What is then an electron? The Danish physicist Niels Bohr said that the question has no meaning. Physics is not about what is but about what we can measure and what we can describe. Both of these concepts, the wave and the particle, are equally correct in describing the nature of the electron.

It isn't just velocity and position that are linked in this inherent uncertainty of nature. If the spin of a particle is known, the orientation of its axis of rotation cannot be defined with precision. Energy and time are similarly linked.

Feeling the Force

The four fundamental forces are interactions between particles. But, how do particles interact? Our image of an interaction is that of a collision. However, a collision between two macroscopic objects involves only the electromagnetic interaction. When two balls collide, the atoms that make up the balls never come into direct contact with each other. These atoms are held together by strong electrostatic attractions; that is, by the electrical attraction of the positively charged nuclei of the atoms and their corresponding electrons. These collisions are ultimately electrostatic interactions.

What then is a force? How does an electron, for example, know that there is another electron in its vicinity so that it can repel it? How does the Earth know that the moon is 385,000 km (240,000 mi) away? If you orient two magnets so that their opposite poles face each other, why do they increase their attractive force as you bring them closer? Clearly, there has to be some signal that is exchanged between the two bodies.

Our modern view of the action of a force originates with the development of the relativistic quantum theory of the electron in 1928 by P. A. M. Dirac, completed later on by Richard Feynman, Sin-Itiro Tomonaga, and Julian Schwinger. According to Maxwell's electrodynamics, when an electron accelerates, it emits electromagnetic radiation. Quantum mechanics tells us that electromagnetic radiation consists of photons, which means that the acceleration of an electron results in the emission of one or several photons. In *quantum electrodynamics*, or QED—the theory that Feynman, Tomonaga, and Schwinger developed—when an electron approaches another electron, the electrostatic repulsion between them is due to the interchange of photons between the electrons.

Where do the photons exchanged by the interacting electrons come from? The answer is that they come from nowhere. They are allowed to exist by the uncertainty between energy and time. If the energy of the interacting electron is uncertain by a very small amount and our knowledge of the time during which the electron has that energy is uncertain by another small amount, during that small interval of time, that small amount of energy can be used to create the photon. As long as the photon is absorbed before the small interval of time is up, no experiment can detect the missing energy from the electron. These particles that can be created for only the time allowed by the uncertainty principle are called *virtual* particles.

Virtual photons are the messengers exchanged by interacting electrons to respond to each other's presence. This continuous exchange of virtual photons constitutes the electromagnetic force that they feel. It turns out that all of the four fundamental forces are the result of the exchange of a virtual particle.

An unexpected consequence of Dirac's work on his version of the relativistic quantum field theory was that his equations gave him two solutions for the energy of an electron, one positive and the other one negative. Dirac interpreted the negative solution in what at the time was an extraordinarily imaginative way. He proposed that the negative energy solution described a particle exactly like the electron but with a positive electric charge. He called the particle the *antielectron*. His bold prediction seems to have been made more out of desperation than foresight. "I didn't see any chance of making further progress," he said later. "I thought it was rather sick."[3] His antielectrons, or *positrons*, as they were named later, were discovered four years later in cosmic rays.

A more detailed examination of Dirac's theory told physicists that relativistic quantum field theory predicted the existence of other *antiparticles* and that for every particle there should exist a corresponding antiparticle. Today, many antiparticles have been discovered in the laboratory and the existence of *antimatter* is well established.

THE BUILDING BLOCKS OF MATTER

Quantum mechanics gives us the rules that govern the behavior of particles and the interaction of particles with other particles through the four forces of nature. But how is the universe put together? We think we have a pretty good idea now. Throughout the twentieth century, physicists painstakingly discovered one by one how the constituents of our world are put together.

The picture that emerged during the early twentieth century was that atoms are composed of dense, positively charged nuclei that are surrounded by the negatively charged electrons. The nucleus contains 99.95 percent of the mass of the atom and has dimensions of the order of 10^{-10} m. The nucleus itself has dimensions of about 10^{-15} m. Soon, physicists realized that electrons were elementary particles, meaning that they were not composed of anything. Their electric charge was also found to be fundamental.[4] Electrons were true building blocks of matter. But the protons and the neutrons that make up the nucleus were not. By the mid-twentieth century, scientists had discovered many more particles that showed complexity and therefore could not be elementary.

In an attempt to make sense of the many new particles being discovered, physicists classified the particles into two large groups. Those particles that felt the strong force were given the generic name *hadrons*, from the Greek *hadron*, or heavy. Particles that didn't participate in the strong interaction were called *leptons*, from the Greek *leptos*, or light. Leptons participate in the weak interaction. Protons and neutrons were hadrons, whereas an electron was one of the leptons.

In the 1950s, a new type of particle was produced via the strong interaction but decayed through the weak interaction. Because of its strange behavior, this new particle was called the *strange* particle.

Although the classification into hadrons and leptons helped, new experiments kept generating new particles with no end in sight. Some scientists wondered if nature was really that complicated or if they were actually missing some important clue.

The clue would come soon. In 1961 Murray Gell-Mann of the California Institute of Technology and, independently, George Zweig, also of Caltech, suggested that all the hadrons were composed of members of a fundamental triplet that Gell-Mann called *quarks*. The odd name started out as "quork," a term he came up with for the peculiar particles with fractional electric charge that he was predicting. A few months later, Gell-Mann found himself reading James Joyce's *Finnegans Wake* and came across "Three quarks for Muster Mark." Quark sounded better than quork! And Joyce had mentioned three, just like the three particles that he was proposing. Quarks they became.

The three quarks that Gell-Mann originally proposed are now called *up*, *down*, and *strange* (or *u*, *d*, and *s*) quarks. These three varieties are known as *flavors*, a term coming from the early use of names like "strawberry," "vanilla," and "chocolate" for these particles. Quarks have fractional electric charge, a finding that initially worried Gell-Mann, since the charge of the electron was supposed to be the fundamental charge and his theory was predicting that there were charges smaller than the fundamental charge. As a result, he initially didn't think that quarks really existed.

As it turned out, the prediction was correct. Hadrons are made up of quarks. Particles like the proton and the neutron are combinations of three quarks. In the case of the proton, the fractional charge of each one of the three quarks adds up to one (in units of the fundamental charge). The total charge of the three quarks that make up a neutron is zero, since the neutron is electrically neutral.

Thus, quarks combine to form hadrons, but the way this works is not simply by choosing quarks until they match their charges. Rather, the mathematical rules of symmetry provide the mechanism that governs the combination of quarks to form each particle.

In 1973 Sheldon Glashow of Harvard University extended Gell-Mann's theory of quarks using mathematical symmetries. The more comprehensive theory revealed the existence of a fourth quark, which Glashow initially named the *charm* or *c* quark, for "the symmetry it brought to the subnuclear world."[5] Unlike Gell-Mann, who originally didn't think his mathematical construct was real, Glashow offered to eat his hat if the charm quark wasn't found experimentally within two years.

I don't know whether Glashow owned a hat then, but if he did, he didn't have to eat it. In November of 1974, two experimental teams found the charm quark simultaneously. One team was led by Samuel Ting at Brookhaven National Laboratory in Long Island, New York, and the other team was led by Burton Richter at the Stanford Linear Accelerator Laboratory in California. Ting, who is of Chinese descent, named the particle the *J* because his name in Chinese begins with a character that resembles the letter *J*. Richter named it the Ψ (pronounced "psi"). The particle is now known as the J/Ψ.

Soon after the discovery of the *c* quark, two other quarks were proposed: the *bottom* or *b* and the *top* or *t*. The bottom quark was discovered experimentally in 1977 at Fermilab and the top quark was discovered in 1995 also at Fermilab, after a long search. By then, physicists had discovered the existence of six leptons: the electron, the muon, the tau, and their associated neutrinos. The electron neutrino, ν_e,[6] is involved in the weak interaction involving the neutron. The muon neutrino, ν_μ, is involved in the decay of the muon. Similarly, the ν_τ is involved in the decay of the tau lepton. These sets of quarks and leptons were classified in three families or generations (Table 5–1).

Are there more generations of particles that we haven't discovered yet? Experiments that were performed as far back as 1991 indicate that these three are all we get. The measured abundances of the light elements present in the universe and the production of certain observed particles are consistent with the existence of three generations.

Table 5–1. Quark and lepton classification into three families or generations

	Quarks		Leptons	
First Generation	u	d	e	ν_e
Second Generation	c	s	μ	ν_μ
Third Generation	t	b	τ	ν_τ

So, we are pretty sure that there are only three generations of particles. But here comes something perplexing. As far as we know today, the observable universe is constructed with the particles of the first generation: the electron, the muon, and the up and down quarks. All the stars, planets, and atoms that we see in the world today are made up of these particles. But nature provides us with two more generations of particles that we don't see the use of. It's clear that we don't have a full understanding of the way nature works.

Symmetry

The four fundamental forces of nature together with the fundamental particles and the rules of quantum mechanics that govern how particles interact with each other via the exchange of virtual particles constitute the laws of physics, the watchmaker of the universe. The laws dictate everything that takes place anywhere in the universe. Were these laws present at the beginning of the universe? Did they create the universe?

While we hope to answer these questions in the chapters ahead, we do know with a great deal of confidence that if the laws of physics existed at the beginning, they did not exist in their present form. The rules have evolved into what they are today. Although remarkably simplified, the laws of physics today, with the four fundamental interactions and the set of particles, are still complex. According to our current view of the laws of physics, during the extremely high temperature and density of the very early universe, before the 10^{-43} second after the big bang, the universe was in a highly symmetric state with only a single unified force in existence. This single force encompassed all the forces operating in our universe today. As the universe evolved and the temperature and density decreased, this original superforce gave rise to the four forces that govern our present-day universe. Running the clock of the evolution of the universe backward, physicists have been able to reproduce the process of unification of the forces from the state of broken symmetry that exists today to the symmetric state that existed before 10^{-43} second. Since the conditions of the very early universe are so extreme, this ambitious program has been done theoretically, with pen and paper. However, in high-energy accelerators in Europe and in the United States, the conditions of the early universe have been partially reproduced, and the observations confirm many of the theoretical predictions.

The process of unification of the forces of nature started in the mid-nineteenth century, when James Clerk Maxwell unified what until then had been two separate forces: electricity and magnetism. Extending and synthesizing the work of Ampère and Faraday, Maxwell unified the two forces guided by the symmetry in electric and magnetic phenomena. A modern view of Maxwell's theory shows that the unified theory of electromagnetism has a more powerful type of symmetry known as a *local* symmetry. The idea of local symmetry was introduced in 1818 by the physicist Hermann Weyl, who was attempting to unify electromagnetism and general relativity. He had been inspired by the work of the German mathematician Amalie Emmy Nöther, who had introduced a powerful theorem that states that every continuous symmetry in the laws of physics is connected with a quantity that remains constant under some transformation.

Recall that a system is symmetric under a certain operation when it remains unchanged under that operation. Rotating a cylinder about its longitudinal axis by any angle keeps the cylinder unchanged. Weyl proposed that the unification of electromagnetism with general relativity would yield a theory that would remain constant or invariant under any space translations or contractions. However, if you translate an object from one location to another, you change the system. If the dimensions of an object shrink, you notice immediately. Weyl's symmetry required that the standards of length and time change at every point in space and time. He called the new invariance *gauge invariance*, because it reminded him of the gauge block used to maintain the distance between railroad tracks constant.

Weyl's gauge invariance is *local* because the standards of length and time change at every point. In a *global* gauge invariance, the symmetric transformation is the same everywhere at once. As an example of a global symmetry, suppose that suddenly the dimensions of everything in the world were reduced by half. The room you're in and all the objects, your house, and the entire town would be now half of what they were a moment ago. Your body would also undergo this immediate and sudden transformation. The same would happen to everyone on Earth and at the same time. Even the Earth would be now half of what it was. The solar system and the visible universe would be reduced by half. If the transformation were immediate and took place everywhere at once, no one could tell the difference. If in addition all the measuring devices and instruments were reduced by half as were all the physical processes and constants of nature, it would be impossible for anyone anywhere to detect the change.

If the size change took place only in certain parts of the world, the change could easily be detected. How can there be local symmetries then? How can we have undetected changes in some parts of a system or at different times and still maintain the entire system unchanged? By introducing a compensating change, the system can remain symmetric. When you watch a car race on television, the images of the cars change size on your screen as the cars race around the track. When the cameraman follows one particular racing car around the track, he zooms in and out to fill the screen with the car. The zooming action is the compensating mechanism that keeps the size of the image unchanged. In this case then, as the car goes around the track, repeatedly approaching and retreating from the camera location, the symmetry of the image on the screen can be restored.

THE HIGGS MECHANISM

Electricity and magnetism individually exhibit global symmetry. But electromagnetism, which unifies the two interactions, exhibits local symmetry. In 1954, it occurred to the physicist C. N. Yang and his collaborator Robert Mills that the local gauge transformations in electromagnetism should be extended to other forces. To do that, they introduced a set of new fields that described a family of particles. Two of the fields turned out to be the electric and the magnetic fields that described the photon. The other fields described photons that were electrically charged, one positively and the other one negatively.

Charged photons? Nothing like that had ever been observed. Could they nevertheless exist even if no one had seen them before? New particles are at times predicted by physics theories and only later turn up in experiments. Would this one be such a case? The problem with charged photons is that they would have to be massless, and massless, electrically charged particles would be produced in very large quantities with any source of electrical charges or magnetic fields. Such an abundance of charged photons would neutralize electric fields everywhere, which is clearly not what we experience. As a result, the Yang-Mills model was shelved since it didn't seem to apply to reality.

But not everyone forgot about the Yang-Mills model. Richard Feynman at Caltech decided to look at the possibility of unifying quantum mechanics with general relativity using the Yang-Mills formalism. While he didn't succeed—no one has—in the process, he did introduce the concept of virtual particles.

Martinus Veltman of the University of Utrecht thought that since there were several significant things in common with the weak interaction and electromagnetism, an attempt at unifying the two forces using the Yang-Mills formalism was needed. The problem was that the weak interaction was short range and the Yang-Mills charged photons were massless. It was known then that a short-range force is mediated by a particle with mass; while a force that acts throughout the universe, such as the electromagnetic force, is mediated by massless particles, like regular photons. Since the weak interaction was short range, it required a particle with mass that the original Yang-Mills model didn't provide. Would it be possible to take the Yang-Mills equations and modify them so that the charged photons end up with mass?

Could the solution be as simple as adding mass to the particles in the equations? Few things are that easy, and this case was no exception. The problem with adding mass by hand to the equations is that in a quantum theory, any arbitrary thing you attempt to do to the equations creates havoc with the theory. In this case, it starts to predict infinite values for many commonly measured quantities, a problem known as making the theory "not renormalizable." In short, not plausible.

The correct solution came from several physicists at about the same time. Higgs's contribution was indirect but fundamental. Higgs, a professor at the University of Edinburgh, had been inspired to study theoretical physics when he became aware of a famous former student at his grade school, Paul Dirac. In a short paper published in 1964, Peter Higgs demonstrated that the spontaneous symmetry-breaking mechanism generated particles with mass, an issue that was being argued in the physics literature at the time.[7] Recall from chapter 4 that spontaneous symmetry breaking is the phase transition that occurs when a system passes from a symmetric but unstable configuration to a stable but no longer symmetric configuration, as when a pencil balanced on its tip topples. In a follow-up paper, Higgs tried out his idea with the simplest gauge theory of all, electrodynamics. The paper arrived at the journal on the same day that a published paper by François Englert of the Free University of Brussels and Robert Brout of Cornell University appeared, which had reached the same conclusions.[8]

According to the method that Higgs and, independently, Englert and Brout proposed, the masses of particles could be altered when they interact with a new field. As the particles move through the field, they gain mass in proportion to the strength of their interaction with the field. This field is known today as the *Higgs field*, and is the same field that I discussed in chapter 4 and that plays a central role in the inflationary theory of the universe. In quantum mechanics, each field produces its own particle. The Higgs field is the source of Higgs particles.

The Higgs field has an unexpected property. For all other fields in physics, the energy takes its minimum value when the field vanishes everywhere. The state is called the *vacuum*. In the case of the Higgs field, however, the energy takes a minimum value when the field has a nonzero value and doesn't vanish in the vacuum. It's similar to what happens if you balance a marble on the tip of the Mexican hat shape we used in chapter 4 (Fig. 5-5). The hat represents the energy density of the Higgs field. The distance from any point on the hat to a vertical axis passing through the tip is the value of the field at that point. When the marble is balanced on the tip, where the field is zero, it has some energy, since it can roll down from this hill. If you disturb the marble slightly, it rolls off, swinging back and forth along the bottom part of the hat's brim, and eventually stops at the lowest part of the hat. At this location, the field has a nonzero value and the energy of the ball has a minimum value since it can't roll down any farther.

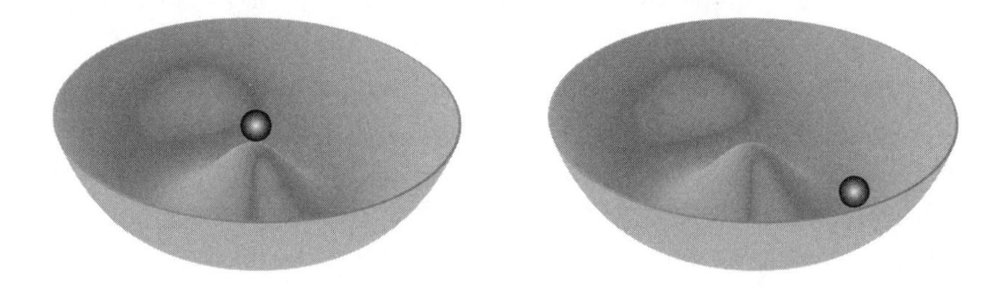

Figure 5-5. A Mexican hat shape representing the energy of the Higgs field. The value of the Higgs field at every point on the hat is the distance from that location to a vertical axis passing through the center of the hat. A marble balanced on the tip of the hat has a zero Higgs field but a nonzero value of the energy since it can roll down. When the ball is at the bottom of the hat's brim, the field has a nonzero value but the energy has its minimum value.

When the marble is at the top of the Mexican hat, where the Higgs field is zero, the configuration is symmetric under rotation around the vertical axis passing through the center of the hat. However, when the marble rolls off to the bottom of the hat, where the Higgs field acquires a value, the configuration is no longer symmetric. The symmetry is spontaneously broken. This phenomenon is called the *Higgs mechanism*.

The Electroweak Unification

Higgs considered only the electromagnetic field. In 1967, Abdus Salam of Imperial College London and Steven Weinberg of Harvard University independently developed a theory in which the Yang-Mills model is responsible for the weak interaction and in which the Higgs mechanism gives mass to the charged photons—which could now be the mediators of the weak interaction.

Salam and Weinberg came from opposite ends of the world. Salam was born and raised in what is now Pakistan and studied at Punjab University. In 1946 he won a scholarship to Cambridge University, where he obtained undergraduate degrees in physics and mathematics. He later received his PhD in physics from Cambridge and returned to Pakistan in 1951 to teach mathematics at Punjab University. Wanting to pursue a career as a researcher in theoretical physics, he decided to return to England, where he became professor of physics at Imperial College.

While at Imperial College, Salam decided that since the Yang-Mills idea was aesthetically pleasing, it had to be the right approach to a model of the weak force. When he attempted to build his model, he realized that a system that is not symmetric can be described by symmetrical equations. Salam's idea was actually the concept of spontaneous symmetry breaking that Heisenberg had introduced in 1928 and that we discussed in chapter 4 in connection with the inflationary theory of the universe.

Steven Weinberg was born in 1933 in New York City. At the age of fifteen or sixteen, when most American students don't even know what physics is, Weinberg had already decided to become a theoretical physicist. He obtained his undergraduate degree from Cornell and his doctorate in physics from Princeton. After graduation, he worked at Columbia, UC Berkeley, MIT, Harvard, and finally, the University of Texas. While at MIT, Weinberg's interests turned to broken symmetries and the unification of the weak and electromagnetic interactions.

In 1967, Weinberg proposed that the Yang-Mills massless charged photons could be the mediators of the weak interaction if one made use of the spontaneous symmetry-breaking mechanism. A couple of months later Salam would reach the same conclusion. According to their theory, the mediators of the weak interaction are massless. At low energies, these particles acquire mass through the Higgs mechanism of spontaneous symmetry breaking. Weinberg believed that besides the photon, his theory predicted the existence of three other particles: one with positive charge, one with negative charge, and the third one neutral. The Higgs mechanism, causing the system to fall to the lower, stable energy state, gives masses to three of the four particles. There was general agreement that the three massive particles should be the mediators of the weak interaction, a short-range force. The three particles were called the W^+, the W^-, and the Z^0. The fourth Yang-Mills particle was the ordinary photon, mediator of the electromagnetic interaction.

Down the hall from Weinberg's office at Harvard was the office of his former classmate Sheldon Glashow. They had both attended the Bronx High School of Science in New York City and Cornell University. Glashow and his advisor, Julian Schwinger, had realized as early as 1956 that the weak and electromagnetic interactions should be unified into a gauge theory. Glashow went on do just that after obtaining his doctorate in 1958. His theory predicted four Yang-Mills particles, two of them charged and two neutral. At the time, the Higgs mechanism hadn't been discovered and there was no way of giving masses to the Yang-Mills particles, so his theory had to be abandoned. But his effort led to the unveiling of the correct mathematical technique that Weinberg and Salam later successfully used.

Electromagnetism and the weak interaction are both gauge theories and their forces are mediated by the same family of particles, the photon, the *W*s, and the *Z* particles. These two interactions are very different today because at the low energies of the present universe, the Higgs mechanism has broken. The Weinberg-Salam theory predicted that at extremely high energies, the symmetry would be restored, and the Higgs field would return to the vacuum, at the top of the hill of the Mexican hat, acquiring a nonzero value. In this state, the four mediators would be massless and the two forces, the electromagnetic and the weak force, would become one: the electroweak force.

The Weinberg-Salam electroweak theory successfully showed us how to use the Higgs mechanism to generate the masses of the weak force mediators. But was the theory renormalizable—that is, did it avoid giving infinite values for certain measured quantities? Because if the theory turned out to be not renormalizable, it would have to be discarded, since those specific quantities were known to be finite. Answering that question proved not to be an easy task. However, in 1971, four years after the introduction of the theory, Gerard 't Hooft and his former PhD thesis advisor, Martinus Veltman, of the University of Utrecht, succeeded in mathematically demonstrating the renormalizability of the electroweak theory.

One thing remained to be done: to see if the particles exist. No one had observed them before, but in 1983, an international team of hundreds of physicists led by Carlo Rubbia, working with an exquisitely precise and sophisticated experimental setup designed by Simon van der Meer at CERN, the European Center for Nuclear Research, found uncontroversial evidence of the existence of the *W* and the *Z* particles. This finding confirmed the prediction of the electroweak theory. The Nobel committee showered the main characters in the saga of the electroweak theory with the physics prize, starting with Weinberg, Salam, and Glashow in 1979, followed by Rubbia and van der Meer in 1984, and finally with 't Hooft and Veltman in 1999.

The Color Gauge Symmetry

After the successful development of the electroweak theory, physicists went to work on the next obvious step, the unification of the electroweak force with the strong interaction. The strong interaction between protons and neutrons, which holds together the nuclei of atoms, is actually the residual interaction of the forces between quarks. As noted earlier, quarks have frictional electric charge and combine to form hadrons of integer charge. The proton, for example, is made up of two *u* quarks each with charge +2/3 (in units of the elementary charge *e*, the charge of the electron) and a *d* quark with charge –1/3, adding up to +1.

The theory that explains how quarks interact to form hadrons is called *quantum chromodynamics*, or QCD, which was modeled after quantum electrodynamics. According to QCD, quarks have an additional charge called *color charge*. In quantum electrodynamics, or QED, there are two kinds of electric charge, which are called positive and negative, although the names have nothing to do with negative or positive numbers. The quark interaction is more complicated than the electromagnetic interaction and the charge comes in three varieties or colors. Color charge has nothing to do with the colors of the visible spectrum, except that the mathematical rules used in QCD for the behavior of these charges have some similarity to the way the three primary colors combine to form white. Each one of the six quarks has three color charges: *red*, *green*, and *blue*, while the antimatter quarks, or antiquarks, have the corresponding *anticolors*: *cyan*, *magenta*, and *yellow*, which combine like those on the color wheel.

A proton has a red *u* quark, a blue *u* quark, and a green *d* quark (Fig. 5-6). These three color charges combine to give a *white* or neutral color proton in the same way that the three primary colors red, green, and blue together combine to produce white. There are hadrons that are formed by the combination of two quarks. These particles are called *mesons* and are also color balanced with the combination of a quark and an antiquark.

These QCD rules explain the properties of the matter particles that we observe in the laboratory and provide an insight into the theory of strong interactions. QCD is a gauge theory and its local gauge symmetry is the invariance of the local transformation of color. Since the symmetry is local, the change in the color charge of one quark can take place while the other quarks that form the particle remain unmodified. The overall effect of the change in color of a single quark in a particle would be to give that particle color. As we saw previously, that change must be accompanied by a compensating change that maintains the symmetry, so that after the transformation takes place, we end up with a white particle.

As with the electroweak theory, the way to accomplish this compensation and balance the change is by introducing new fields. The messenger particles of these fields are called *gluons*. There are eight gluons, one for each compensating field. As messenger particles, the gluons play the same role as the photon in electromagnetism or the *W* and *Z* particles in the electroweak theory: they transmit the interaction.

Gluons are exchanged by quarks much in the same way that photons are exchanged by electrons. But the photons that transmit the electromagnetic force between electrons and other charged objects don't carry charge themselves. Gluons do. A gluon carries one color and one anticolor, for example, a red and an antiblue. When we perform a local gauge transformation that starts with a red quark that forms part of a proton and turn it into a blue quark, a red/yellow gluon is created simultaneously. Since yellow is the anticolor of blue, yellow and blue neutralize each other. The net color that results is blue + red + yellow = red, and the symmetry is restored.

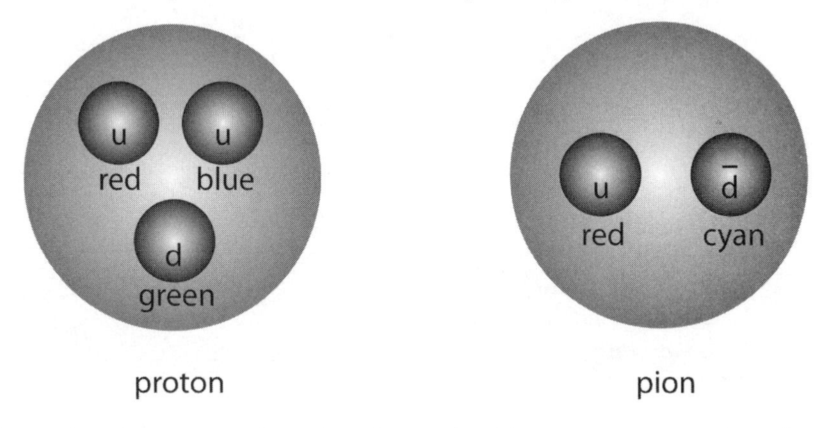

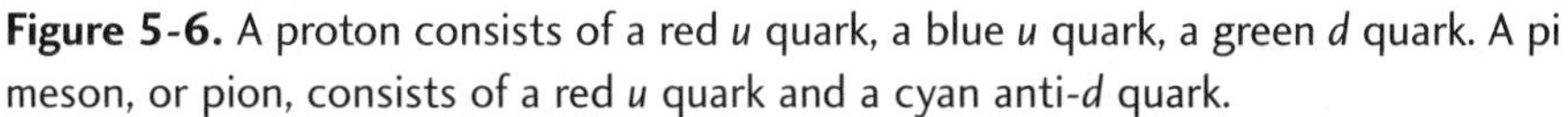

Figure 5-6. A proton consists of a red *u* quark, a blue *u* quark, a green *d* quark. A pi meson, or pion, consists of a red *u* quark and a cyan anti-*d* quark.

The Standard Model

The description that I have just given—of the four fundamental interactions, the messenger particles that transmit these interactions, and the quarks and the leptons, the elementary matter particles that make up the universe—constitutes what is known as the *Standard Model* of particles and forces. The matter particles have a quantum mechanical property that resembles mathematically the spin of a ball. Quarks and leptons can be thought of as tiny gyroscopes, with one fundamental difference from an actual gyroscope. The *spin* of an elementary particle is quantized and can only take values that are multiples of Planck's constant divided by the quantity 2π. In terms of this unit, a particle's spin has to be either an integer (including zero) or an integer plus 1/2. For each type of particle, the total spin is fixed, although its orientation is not. Particles with half integer spin are known as *fermions*, in honor of Enrico Fermi, who introduced the statistical treatment that describes their behavior. Integer spin particles are called *bosons*, in honor of

Table 5–2. The Standard Model of Matter Particles and Force Carriers

		Fermions			*Bosons*	
Matter Particles	**Quarks**	u up	c charm	t top	γ photon	**Force Carriers**
		d down	s strange	b bottom	g gluon	
	Leptons	ν_e electron neutrino	ν_μ muon neutrino	ν_τ tau neutrino	Z *Z* boson	
		e electron	μ muon	τ tau	W *W* boson	
	Generations	**I**	**II**	**III**		

Saryendra Bose who, along with Einstein, developed the statistical quantum-mechanical description of these particles.

The quarks and the leptons listed in Table 5–1 are fermions. A list of the Standard Model matter particles (fermions) and the force mediators (bosons) for the three generations appears in Table 5–2.

GRAND UNIFICATION

Since QCD is a gauge theory it can in principle be unified with the electroweak theory, which also is a gauge theory, as we have seen. That's what Howard Georgi and Sheldon Glashow of Harvard University proposed in a 1974 paper.[9] According to the Grand Unified Theory—or GUT, as they called their model—during the very early universe, at an earlier time than when the electroweak unification took place, when the temperature was a hundred trillion trillion kelvin, electromagnetism and the weak force along with the strong force would have been unified into a single superforce. As the universe evolved and cooled, the Higgs mechanism broke the first symmetry and the strong force split off from the electroweak force. An instant later, when the universe cooled even more, the electroweak force split into the electromagnetic and weak forces through a second symmetry-breaking mechanism.

In our cold world, the three forces are very different. They grew apart. The Higgs mechanism differentiated them from each other as the temperature of the universe dropped. But moments after the birth of the universe, they were one and the same, a single parent force.

A theory that unified the three nongravitational forces would treat quarks and leptons on an equal footing. The single unified force would allow a quark to change into a lepton and if this were to happen, the number of quarks in the universe would change. If the quarks inside the proton changed into leptons, the proton would decay and all matter in the universe would be unstable. Physicists calculated that, if the GUT were correct, the lifetime of a proton would be 10^{30} years, a time span many orders of magnitude larger than the age of the universe!

Since the universe has been around for 14 billion years, which is "only" about 10^{10} years, there seems to be no hope of catching the decay of a proton to test the theory. However, since the lifetime is an average value, some protons can decay much faster than 10^{30} years while others take much longer. It's the same with people, which is why insurance companies can stay in business while offering policies of, say, one hundred thousand dollars to a thirty-year-old man who would pay only a few hundred dollars a year for the policy. The insurance companies know that the average life span for men in an industrialized country is seventy-six years and that most of their insured will live long enough to pay for both their policies and those of the few who die earlier. If you consider only two or three men, it's unlikely that one of the three will die in his infancy, but if you look at the life spans of a million men, you'll find many who die when they are months old. If the lifetime of a proton is 10^{30} years, you could gather 10^{30} protons together and use sensors expecting to watch at least one decay in a year, which is exactly what physicists did after the decay of the proton was predicted by Georgi and Glashow's theory. Since you can find lots of protons in water,[10] scientists set up elaborate experiments in large water pools deep underground, like the Irvine/Michigan/Brookhaven experiment in the Homestake Mine in South Dakota or the one under the Gran Sasso Mountain in central Italy. Other such experiments were set up in Japan and India. Nonetheless, no one has ever detected proton decay and the estimates are that the lifetime of a proton is much larger than what the GUT theory predicts.

The failure to detect proton decay together with more accurate measurements of the strength of the forces indicate that the Grand Unified Theory of Georgi and Glashow is very likely incorrect. Does this mean that there is no hope for the unification of all nongravitational forces? Not necessarily. The predictions of the Standard Model have been proven to be correct and the theory is very strong. An extension of the Standard Model is one of several possible avenues toward unification.

THE WATCHMAKER

The successful development of the Standard Model has led to an impressive simplification of all the nongravitational forces of nature. The behavior of the handful of building blocks of matter, the elementary particles listed in Table 5–2, is fairly well understood. The six quarks and six leptons that interact with each other by the interchange of four charge carriers make up all the matter in our world today. The symmetric principles that we have discovered show us that the three nongravitational forces at play today come from two main forces that existed briefly during the early universe. These powerful symmetries indicate to us that these two remaining forces should merge into one, a superforce that existed at an even earlier time after the birth of the universe and that gave rise to the three forces that we observe. That our first attempt to discover this superforce didn't turn out to be correct hasn't deterred physicists from continuing to pursue it.

The success of the Standard Model in unifying the electromagnetic force with the weak nuclear force seems to indicate that the next level of unification will be possible as an extension of the model. If and when this unification is achieved, it is likely that it will still imply the decay of the proton, only with a longer lifetime. There is already evidence that the lifetime of the proton should be longer and that the experiments that have been trying to detect it for over two decades shouldn't have detected anything yet.

There are other approaches at unification, such as the supersymmetric models and the multidimensional superstring models. Some of these models unify *all* forces of nature, including gravity, a particularly exciting feature. We will discuss these models in chapter 6.

As we've just seen, the power of symmetry and of the Higgs mechanism have made possible the development of the Standard Model, achieving a simplification of the forces of nature and providing us with an unprecedented understanding of the laws of physics. As I stated earlier, the laws of physics are the watchmaker of the universe. They make the world go around. At the beginning of time, the laws were simpler, and we have seen hints at the idea that there was only a single force at play. These laws controlled the evolution of the universe, the forces of nature, and the way these forces evolved and operated. Everything that happens in the universe happens because the laws of physics allow it to happen. The watchmaker governs the entire universe and not only its evolution, but its own evolution. The question that remains is, did this watchmaker make itself, exist forever, or was it created? This is a question to be addressed in the remaining chapters.

Chapter 6

Is God in the Details?

The Hierarchy Problem

The Standard Model—perhaps the most complete physics theory ever developed—gives us a complete description of the nongravitational particles and forces in the universe. It describes with extraordinary accuracy the way these forces shape our world through the exchange of particles called bosons that act as force carriers. The Standard Model successfully unifies the electromagnetic and weak interactions into the single electroweak force, providing hope for a grand unification of the three nongravitational forces. However, no one has succeeded in taking this last step. There are indications that the Standard Model, as it stands, may not allow for the Grand Unified Theory and that the model may need to be extended.

The difficulties that the Standard Model runs into are deeper than the failure to achieve grand unification. The model makes use of the Higgs mechanism to equip the various elementary particles with mass. The Higgs mechanism is possible because in the Standard Model the vacuum isn't empty but is instead filled with the Higgs sea, virtual Higgs particles that are allowed to exist by the uncertainty principle of quantum mechanics. These virtual Higgs particles appear from the vacuum and disappear again in a very short time that is controlled by the energy needed to create the particle and the limit imposed by the uncertainty principle. As quarks, leptons, and bosons travel through the vacuum, they interact with these virtual Higgs particles and slow down, as if they were traveling through a dense medium.

We can think of the particles' motion through the vacuum as that of soldiers marching through a muddy field. Some soldiers march in areas that aren't too muddy and slow down a little. Others march in thicker mud where it becomes difficult to move and slow down substantially. As they march, the mud sticks to them and they gain mass. The thicker the mud, the more they sink into it and the larger the mass that they gain. The soldier's mass gain depends on the strength of the interaction with the mud. Similarly for the quarks, leptons, and bosons that move through the sea of virtual particles. As they interact with the virtual particles, they gain mass and are slowed down. The particle's mass depends on the strength of the interaction with this sea of virtual particles. The more that the particle is slowed down, the greater the mass it acquires.

But quarks and leptons aren't the only particles that gain mass from the Higgs sea. The Higgs particle itself also gets its mass from the interaction with the virtual Higgs sea of the vacuum. But there is one important difference. When other particles interact with the virtual Higgs particles, they obtain their masses—and those masses coincide with the measurements that we obtain in our experiments. Not so with the Higgs particle. The Higgs interacts not only with the virtual Higgs particles but also with other particles, many of which are very massive. As a result of the interactions with heavy particles, a new sea of virtual particles—virtual fermions and bosons—and their virtual antiparticles are created due to the quantum fluctuation mechanism allowed by Heisenberg's uncertainty principle. A real Higgs particle acquires its mass from the interactions with the Higgs sea and from interactions with this sea of virtual fermions and bosons. Since some of these particles are heavy, the interaction is strong and the contribution to the real Higgs particle's mass is large. As the Higgs particle moves, it interacts with many particles that contribute to its total mass.

The problem with the Standard Model in this case is that the mass of the Higgs particle should be relatively light if it is to describe correctly the weak interaction. However, the theoretical calculations for the mass of the Higgs particle that take into account all the contributions from the virtual particles that it interacts with give a value that is thirteen orders of magnitude higher than it should be! This problem is known as the *hierarchy problem* and remains unresolved in the Standard Model.

One possible way to resolve it is to assume that all the contributions, large and small, some of which can be negative, all cancel out exactly so that the Higgs mass comes out with the value needed for the weak interaction. According to quantum mechanics, photons, gluons, and, in fact, all bosons contribute positive energy to the vacuum. On the other hand, electrons, neutrinos, and quarks—all fermions—contribute negative energy to the vacuum. The vacuum energy resulting from virtual bosons cancels the virtual energy from virtual fermions. But, is the cancellation total? Is it possible that, if we could account for the contributions from all fermions and all bosons, the total contribution would end up being zero? The only way to know whether this total cancellation can take place is to apply the rules of quantum mechanics to the problem. The answer we get back is that the cancellation is extremely unlikely because the masses of the particles are very different. A Higgs particle encounters virtual electrons, virtual quarks, virtual photons, and all kinds of virtual particles and antiparticles in its path. Some fermions contribute negative energy and some bosons, positive energy. But the negative energy contribution of a virtual electron isn't all cancelled by the positive energy contribution of a virtual gluon or *W* boson. There isn't any single particle with the exact amount of positive energy that can precisely cancel the negative energy of the virtual electron.

However, if you add up the energies of all the fermions surrounding a Higgs particle at any one time, you come up with a certain value. You can now select bosons and additional fermions as needed so that the total sum is zero. It's as if you walk into a large discount store and are given one hundred dollars to spend during a promotion, but with the condition that you must spend it all in the store at once, without taking any cash with you after you're done shopping. If you start with the items you'd want to buy, then, when only a few dollars are left in your pocket, you can take your time looking for inexpensive items that would add up to the total. You might need to shop for these items with calculator in hand and even exchange one of the more expensive items for another of slightly different value until you come out exactly even. You can see that if you were given a much larger sum, say, $50,000, it might be more difficult to do it, especially if the discount store didn't sell any items of over $2,000 or so. Given enough time and determination, you'd likely be able to fine-tune your purchases until you spent every last cent of the $50,000 without spending one cent more. (And on the way home with your loot, it might be wise to stop at the local donation center and make a substantial contribution with the many little trinkets you ended up purchasing just to make the total.)

The fine-tuning needed to achieve the cancellation of virtual negative and positive virtual energies would be considerably much more difficult to achieve. In this case, certain parameters in the Standard Model would have to be fine-tuned to better than one part in a million billion. Although these precise cancellations can be performed in the model, such fine-tuning is difficult to accept with the observed particles. It appears that the Standard Model must be extended with some new physics that can explain the fine-tuning and solve the hierarchy problem.

SUPERSYMMETRY

How could the Standard Model be extended so that it solves the hierarchy problem but still retains the extraordinary agreement with observations that it achieves at the lower energies? Any extension to the model should contain additional particles that interact with the Higgs particle and *naturally* counterbalance the contributions to the mass caused by the known particles in the Standard Model. In a way, the hierarchy problem predicts the existence of these additional particles.

Remarkably enough, a theory with such properties existed even before the hierarchy problem was identified: *supersymmetry*. After the disenchantment with the Grand Unified Theory of Georgi and Glashow when the prediction of proton decay wasn't corroborated, physicists began to look at other possible symmetries. They found hope in supersymmetry, a theory that had been proposed in 1971 by the French physicist Pierre Ramond. His idea was to include fermions in the original version of string theory, which included only bosons.

According to supersymmetry, the laws of physics are symmetric with the exchange of bosons and fermions. Supersymmetry implies that every boson has a fermion *superpartner* with exactly the same properties as the boson except for its spin. If the universe is supersymmetric, then, all particles of nature come in pairs. Recall that fermions, which have half integer spins, are matter particles, while bosons, which have integer spins, are the messenger particles. Supersymmetry pairs matter particles with force particles.

If supersymmetry is correct, it could solve the hierarchy problem in a natural way, since fermions and bosons tend to give quantum mechanical contributions that cancel each other out. If the contribution of a virtual boson is positive, that of the accompanying fermion superpartner is negative and the two contributions cancel out from the start. Higgs particles moving through the sea of virtual bosons and fermions wouldn't receive additional contributions to their masses from the virtual boson-fermion pair, thus allowing the Higgs to remain relatively light, as required by the weak interaction. Supersymmetry provides the natural canceling out of the quantum mechanical contributions of large particles without the fine-tuning that physicists were forced to perform with the Standard Model alone.

All of this is very nice, but which particles are the superpartners? Which boson is the superpartner of the electron, for example? It turns out that none of the Standard Model bosons can be the superpartner of the electron. In fact the superpartners of all the known particles have never been observed. What the theory predicts is that every Standard Model particle has a yet-to-be-found superpartner with the same properties except for its spin. If the Standard Model particle is an electron, which is a fermion with spin 1/2, there must be a superpartner electron, the *selectron*, with the same mass and charge, but with spin 0, which would make it a boson. If the particle is a quark, which has spin 1/2, there must be a spin 0 *squark*. Similarly for all the other fermions, such as the neutrino whose superpartner is the *sneutrino*. The superpartners of the matter particles are *sparticles*.

If the Standard Model particle is a force particle, which is a boson, such as the photon with spin 1, there must be a superpartner photon or *photino*, with zero mass, no electric charge or weak charge or strong charge, that moves at the speed of light, but with spin 1/2, making it a fermion. There must also be *gluinos*, *zinos*, and *winos*, the fermion superpartners of the gluons, and of the Z and W bosons. There should also be a *higgisino* superpartner. If the theory turns out to be correct, we should get used to the new ugly names.

Supersymmetry brings with it an abundance of new particles, none of which has ever been observed. It actually doubles the number of particles in the universe so that the theory is by no means economical. The situation is probably no different from Dirac's 1928 prediction of the existence of the antielectron, which even he didn't think would ever be corroborated. As it turned out, it was corroborated: every matter particle has a corresponding antimatter particle, with exactly the same properties except for its opposite electric charge.

But is nature supersymmetric? To put it another way, if every matter particle and every force particle has a superpartner, where are these particles? Why have we not found even the lightest one?

If superpartners exist but we haven't found them, it's probably because they are too heavy, heavier than our present experiments can detect. But I said earlier that superpartners should have the same mass, charge, and other physical properties as the Standard Model particles, except for spin. What gives? Supersymmetry gives. Our world is not supersymmetric, at least not at the present time. If supersymmetry did exist in the past, it is broken now and the superpartners have much larger masses today than the Standard Model particles. However, the details of the way the symmetry is broken or which particles are involved are not yet known.

PULLING STRINGS

The supersymmetric Standard Model, the extension of the successful Standard Model that includes supersymmetry, explains the hierarchy problem in a natural way, but it does so at a seemingly fantastic cost: it doubles all the Standard Model matter particles and force particles. We should find out within a few years whether superpartners exist with the experiments planned for the new Large Hadron Collider already in operation at CERN, the European Center for Nuclear Research.

Even if the superpartners are found, the supersymmetric Standard Model is not quite the final answer. We still don't have a good theory of the unification of the nongravitational forces, although supersymmetry has brought us closer to achieving it. When Georgi and Glashow first introduced their Grand Unified Theory, their calculations showed that during the early universe, up to 10^{-35} second after the big bang, when the temperature was 10^{28} K, the three nongravitational forces had equal strengths. During that brief interval of time of the early universe, the three forces were unified into a single "superforce." Today, the symmetry is broken and the three interactions have very different strengths. However, more precise calculations of the strengths of the weak and strong forces show that, in the Standard Model, the three forces would not have the same strength at the same time, and the "superforce" didn't exist. But the extension of the model, the supersymmetric Standard Model, might restore the convergence of the forces into a single one (Fig. 6–1). If the energy contributions of the virtual superpartners that are now present are included in the calculations, the convergence is restored and the three forces are again unified at the same time. If and when the superpartners are found and their masses are determined, a new, consistent grand unified theory that replaces Georgi and Glashow's GUT might be developed.

Even then, the extended supersymmetric Standard Model that includes a correct Grand Unified Theory would still not be the final answer since the model in all its reincarnations doesn't include gravity. And it doesn't explain the patterns seen in the masses of the matter particles.

There is, however, a supersymmetric theory that does both: *superstrings*. Plain string theory—without the *super*, since there's nothing "plain" about string theory—was actually invented before supersymmetry was introduced.

String theory started out in 1968 as an effort to explain the abundance of elementary particles known as hadrons that were being detected in great quantities in many laboratories around the world. Hadrons, as you'll recall, are particles that interact via the strong force. Protons and neutrons are hadrons. The puzzling situation at the time was that experimenters were discovering new hadrons just about every time they turned on their instruments. At first, people got excited and began to name these new particles. But soon, the excitement turned into embarrassment, as the number of new particles kept growing and no one could come up with a reasonable classification scheme that could explain their existence. In an attempt at explaining the interactions of all these particles, the Italian physicist Gabriele Veneziano of CERN proposed a precise formula that seemed to do the trick.[1]

A close investigation of Veneziano's formula by Yoichiro Nambu of the University of Chicago and independently by Leonard Susskind of Stanford University and by Holger Nielsen of the Neils Bohr Institute revealed that it represented the interaction between vibrating strings.[2] By assuming that the interacting particles were vibrating strings, the proliferation of new particles could be tamed, since many of those were simply the different quantum states of a vibrating string.

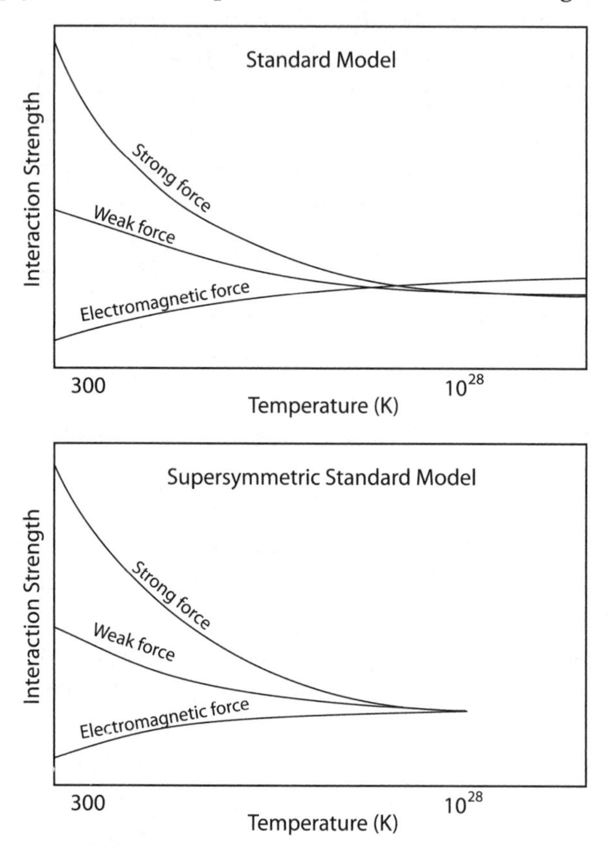

Figure 6-1. Due to symmetry breaking, the strengths of the three nongravitational forces are different from each other today. Examining their history, we see that in the Standard Model (top), the strengths of these three forces approach one another during the early universe but don't quite meet at the same time. In the supersymmetric Standard Model (bottom), the additional contributions of the virtual superpartners change the interactions. Calculations with this model indicate that the three forces were unified when the temperature of the universe was 10^{28} K, which occurred when the universe was less than 10^{-35} second old.

Soon, some mathematical difficulties developed that were related to the probabilities of the outcome of experiments. In a well-behaved theory, the sum of all possible outcomes must equal 100 percent (which is the same as one). For example, if you place ten red balls, ten blue balls, and ten yellow balls into an empty jar, shake it, and ask someone to close her eyes and draw one ball, the probability of picking a red ball is one third, and the probability of picking a blue ball is also one third, as is the probability of picking a yellow one. The sum of the three probabilities is one. In the new theory of hadronic strings, the probabilities weren't adding up to one, which in our example would mean that the ball that your friend picked from the jar wouldn't be red or blue or yellow, which is impossible.

Fortunately, however, a solution came in an expected form. Physicists realized that the probability problem would go away if they assumed that these strings existed in a world of more than four dimensions. Twenty-six dimensions to be exact!

The idea of considering more than our four familiar dimensions (three of space and one of time) in an attempt at unification dates back to 1919, when the German mathematician Theodor Kaluza cast Einstein's field equations of gravitation onto a five-dimensional spacetime. In Kaluza's model, the fifth dimension turned out to be Maxwell's electromagnetism. According to this model, electromagnetism is a form of gravity in the additional dimension. A few years later, the Swedish physicist Oscar Klein extended Kaluza's theory and showed that the fifth dimension was rolled up in a tiny circle with a radius of 10^{-30} cm. The Kaluza-Klein model turned out to be incorrect and electromagnetism had to be first unified with the weak force—a force unknown in their time—before it could be unified with the other forces. Kaluza himself considered his model to be no more than a mathematical trick, an exercise in the unification of two distinct forces.

In twenty-six dimensions, the string model of hadrons had mathematical consistency. The model, however, ran into difficulties when it didn't agree with new experimental results related to the energies of certain particles that were produced during high-energy interactions between hadrons. The new data supported instead Murray Gell-Mann's idea to classify the strongly interacting hadrons in a symmetric mathematical pattern. The idea eventually resulted in the successful model of the quarks as the fundamental particles that make up all hadrons (as described in chapter 5).

The intriguing model of hadrons as vibrating strings was unsuccessful at its original purpose, but the idea of vibrating strings survived. In 1974, John Schwartz of the California Institute of Technology and Joel Scherk of the École Normale Supérieure in Paris proposed that the failure of the string model could be turned into a success if instead of considering it to be a model for hadron interactions only, it were extended to describe a theory that would unify all the fundamental forces of nature, including gravity.[3] Schwartz and Scherk realized that the earlier model already predicted the existence of a spin two particle, which, when it was proposed, was seen as a failure in the theory, since no such particle existed. But in a unified theory, the gravitational force would be described by the exchange of a massless spin two particle, the graviton, the quantum of gravitation. Previous attempts at unifying gravity with quantum mechanics ran into difficulties when calculations for the exchange of the graviton, considered then to be a point particle, yielded infinities. If the graviton is considered to be a string instead of a point particle, the infinities go away.

In its new expanded role, string theory described *all* particles as vibrating strings. The theory described all the matter particles and the force particles in the Standard Model as well as the graviton, the carrier of the gravitational force. As such, string theory would be a theory that unifies *all* the known forces of nature in a single sweep.

Through the efforts of a handful of physicists, the expanded string theory began to take form, eventually involving supersymmetry as a symmetry of the entire spacetime. When that happened, string theory became *superstring* theory.[4] The introduction of supersymmetry into the theory brought to light some mathematical difficulties related to uncontrolled quantum fluctuations that appeared when the theory was cast in twenty-six spacetime dimensions. The difficulties went away when the number of required dimensions was reduced to ten: nine spatial dimensions plus one of time. As with the Kaluza-Klein model, the extra six dimensions are assumed to be rolled up in a tiny circle and are invisible to us. Physicists call this rolling up of the extra dimensions *compactification*.

The birth of the modern theory of superstrings is marked by the publication of a paper by John Schwartz and his collaborator Michael Green of Queen Mary College in London.[5] It showed that the ten-dimensional superstring theory included fermions, bosons, Yang-Mills fields, and gravitons in a mathematically consistent way that didn't generate any infinities, a problem that plagues most modern theories of matter.

According to superstring theory, all elementary particles are actually the oscillations of tiny strings, no longer than 10^{-35} meter (roughly equal to the Planck length). These strings[6] vibrate like a guitar string and each one of its vibrational modes corresponds to a particle. But unlike guitar strings, these strings are not made up of atoms or electrons or quarks. Atoms are made up of elementary particles called electrons and quarks, which are the oscillations of these strings. What are strings made of, then? The question cannot be answered, because strings are the fundamental building blocks of matter, and a building block is always the end of the chain. A classical symphony, for example, is made up of four movements. Each movement is made of musical notes. What's a note made of? Notes are the fundamental building blocks of music and, as such, are the end of the line. There's no answer to that question either.

The intrinsic properties of an elementary particle, such as its electric charge, spin, or color charge, are determined by the vibrational mode of the string. The frequency at which the string vibrates, for example, determines the energy and therefore the mass of the particle (from Einstein's mass-energy equation, $E = mc^2$). An individual string corresponds to several particles, each one a different mode of oscillation of the string. There are several types of strings, each one oscillating in many different modes. There are open strings and closed strings. *Open strings* have two endpoints while *closed strings* are loops.

Open strings can split into two strings and two strings can join the endpoints and become a single string. An open string can also form a loop and break into an open string and a closed string. The two endpoints of an open string can join together to form a closed string. Finally, two open strings can interact with each other, joining and breaking up into two new strings. These possible interactions of open strings were initially identified by Michio Kaku of New York University and Keiji Kikkawa of Osaka University by using string field theory. Closed strings, on the other hand, undergo only one interaction, combining and splitting. The possible interactions of open and closed strings are shown in Fig. 6–2.

Figure 6-2. The possible interactions of open and closed strings. Open strings can undergo five possible interactions: two strings can join endpoints and combine into a single string, a string can split into two new strings, a string can split into an open string and a closed string, a single string can join its two endpoints and become a closed string, and two open strings can join and break up again to form two different strings. A closed string can split into two new closed strings.

If superstring theory is correct, the interactions between the strings are the underlying mechanism of the electromagnetic interactions carried by photons that we observe between electrons and protons in an atom, the interactions among quarks that we observe as the strong interaction, or the interaction carried by the *W* bosons in the weak force. String theory also explains Einstein's theory of gravity—the general theory of relativity. The graviton, the quantum of gravity, is interpreted in superstring theory as the lowest vibrational mode of a string. Calculations in string field theory show that the interactions between these strings give us a quantized general relativity. The string theorist Edward Witten of the Institute for Advanced Studies in Princeton has said that if Einstein hadn't discovered his general relativity, the theory would have come out of superstring theory.

String theory naturally unifies all the forces of nature. Although the unification looks straightforward in the theory, the effort to discover it wasn't. The weak interaction, one of the four forces, presented an early challenge to superstring theory. The weak interaction has one very important property: it is the only force in nature that distinguishes left from right. However, after compactification, string theory treats right and left the same way, apparently excluding the weak interaction on those grounds. The discovery of this symmetry violation, as physicists like to say, dates back to the late 1950s. Up until then, researchers had checked the other three forces—gravity, the electromagnetic force, and the strong force—and found that they all complied with this symmetry. It was easy to assume that nature was mirror symmetric, that the laws of physics are the same when left and right are exchanged.

If the laws of physics are mirror symmetric, you couldn't perform any experiment that would tell you whether you're looking at the world directly or in a mirror. When you look at a top spinning counterclockwise, you can't tell whether you are looking at it directly, in which case it is spinning counterclockwise, or at its image in a mirror, in which case the actual top is spinning clockwise.

Two old friends from college, the physicists T. D. Lee from Columbia University and C. N. Yang from the Institute for Advanced Studies in Princeton, considered this symmetry violation and realized that, since no one had looked into the mirror symmetry violation in weak interactions, an experiment needed to be performed. In a now classic paper,[7] Lee and Yang discussed several possible experiments to be performed. However, both physicists were theoreticians. Someone else needed to perform the experiments. All they could do was publish the paper and wait.

A third physicist, Chien-Shiung Wu from Columbia University, took up the challenge. Madame Wu, as she was known to the physics community, was a superb experimentalist, renowned for the elegance and accuracy of her experiments. She was also one of the world's authorities on the weak interaction. When she read Lee and Yang's paper, she was preparing a trip to the Far East with her husband to commemorate their emigration from China exactly twenty years earlier. After she read the paper, she decided that that work was of outmost importance and canceled her trip, leaving her husband to take it alone.

Madame Wu looked at the radioactive decay of cobalt-60, which decays through the weak interaction. This nucleus, like many others, has an intrinsic spin. The cobalt-60 nucleus decays through beta decay and emits electrons. What Lee and Yang wanted to know was whether there was a preferred direction when the nucleus spun one way or the other. If more particles were emitted in the direction of the nuclear spin, the mirror image would show the opposite: more particles moving in the direction opposite the nuclear spin. There would be a preferred direction and a way to distinguish the mirror world from the real world.

Madame Wu's experiments showed that there was an asymmetry in the directions of the electrons emitted, that there was a preferred direction. Additional experiments by her group and others confirmed it. Today, after much experimental evidence, it is a well-established fact that nature is not mirror symmetric. The weak interaction distinguishes left from right.[8]

The ten-dimensional string theory, with six dimensions rolled up in a tiny circle, is mirror symmetric, treating left and right equally. That presented a major problem for string theory, since, as we just saw, the weak interaction violates mirror symmetry. Regardless of what the theory could achieve in the unification of all the forces, if one experiment contradicted it, the theory would not be viable.

Fortunately there was a solution. In 1985, Phillip Candelas, Cary Horowitz, Andy Strominger, and Edward Witten applied a more thorough compactification method known as a *Calabi-Yau manifold* to the compactification of the unseen dimensions in string theory. Candelas's group found that the compactified spaces of the rolled-up dimensions could be described by a manifold—an intricate mathematical object that had been discovered in 1976 by the mathematicians Eugenio Calabi of the University of Pennsylvania and Shing-Tung Yau of Stanford University. Manifolds make possible the representation of a complicated structure in terms of simpler spaces that are already well understood. An example of a manifold is the surface of a sphere. If you draw a large triangle on the surface of a sphere, the sides will curve along with the surface and the sum of the three internal angles will be

Plate I. Supernova 1987A. Remnant of the self-destruction of a massive star in the Large Magellanic Cloud, a satellite galaxy of the Milky Way. The supernova remnant in the center of the photograph is surrounded by inner and outer rings of material. *[Courtesy NASA/ Space Telescope Science Institute/Hubble Heritage Team]*

Plate II. Star cluster in the Small Magellanic Cloud, a satellite galaxy of the Milky Way. This cluster is located about 200,000 light-years away and is some 65 light-years across. *[Courtesy NASA]*

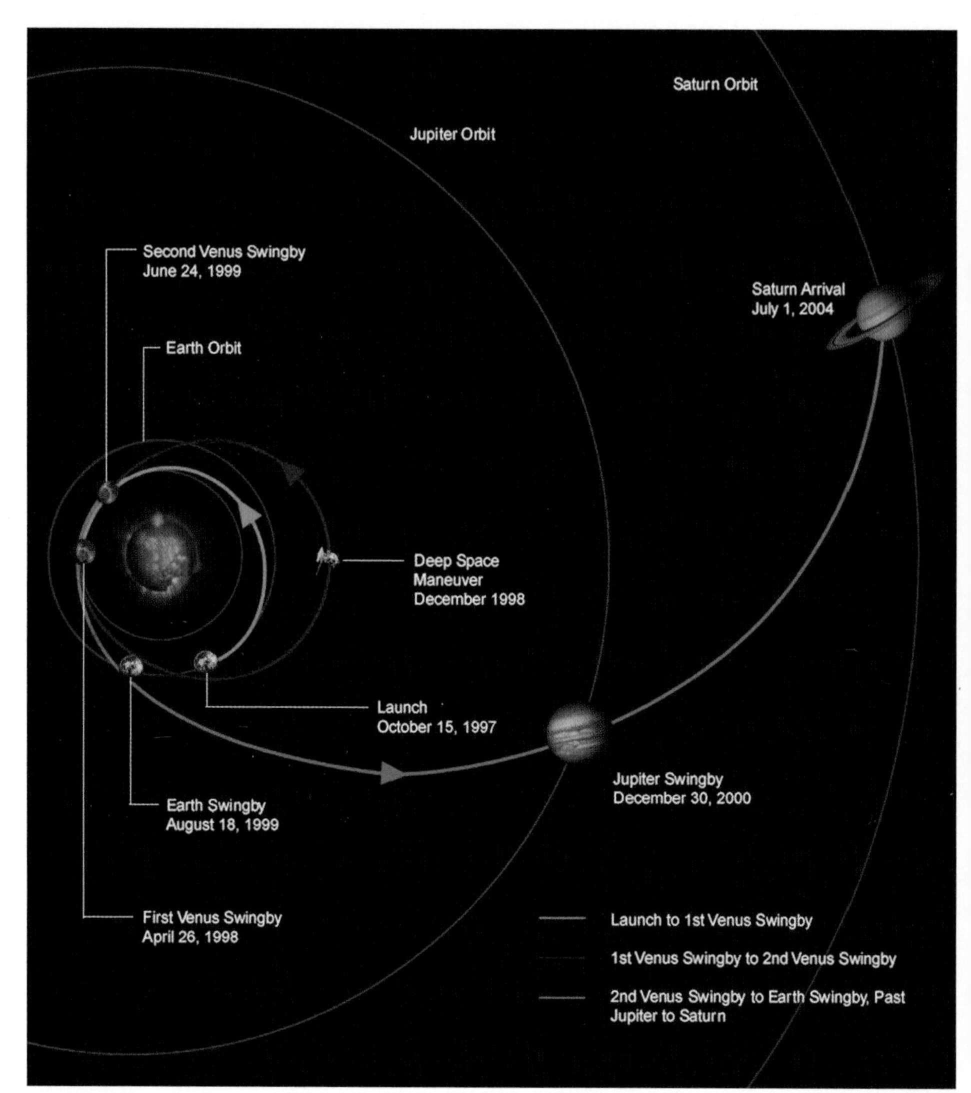

Plate III. The physics of Newton allows us to calculate the complicated interplanetary trajectory of NASA's *Cassini* spacecraft, using the gravitational attraction of Venus, the Earth, and Jupiter to propel the craft toward its final destination around the planet Saturn. *[Courtesy NASA JPL]*

Plate IV. Barred Spiral Galaxy NGC 1300 in the constellation Eridanus, 69 million light-years away. *[Courtesy NASA/Space Telescope Science Institute/Hubble Heritage Team]*

Plate V. Galaxy AM 0644-741, a member of the class of ring galaxies, is located 300 million light-years away in the direction of the southern constellation Voltans. The ring is a shock wave that resulted from a collision with another galaxy. As the ring plows outward, gas clouds collide and are compressed. The compression eventually results in the formation of new stars. *[Courtesy NASA/Space Telescope Science Institute/Hubble Heritage Team]*

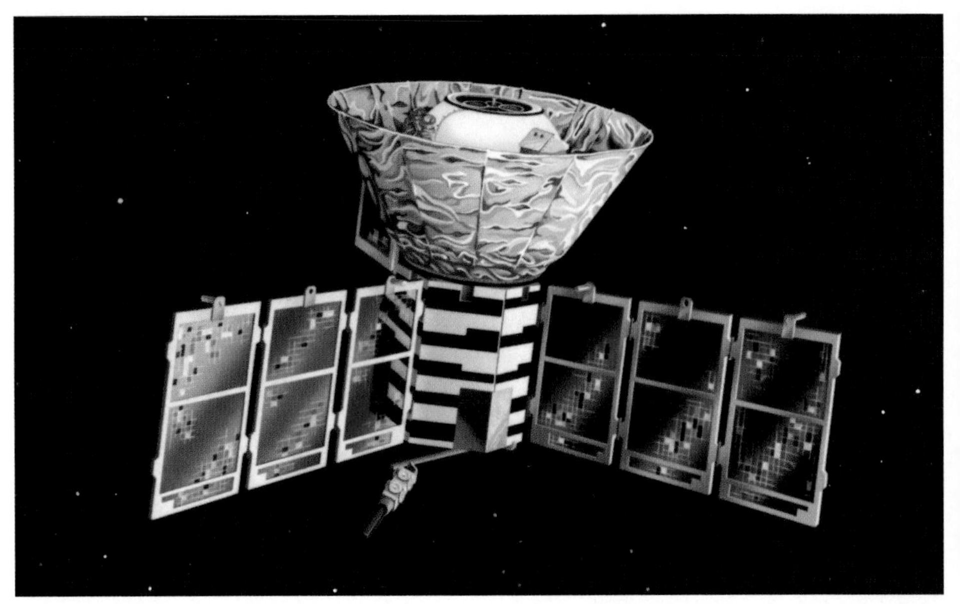

Plate VI. NASA's Cosmic Microwave Background Explorer (COBE), launched in November 1989 to measure the echo of the big bang with unprecedented precision. The satellite contained three instruments: DIRBE, a Diffuse Infrared Background Experiment, to search for cosmic infrared background radiation; DMR, a Differential Microwave Radiometer, to map the cosmic radiation sensitivity; and FIRAS, a Far Infrared Absolute Spectrophotometer, to measure the spectrum of the CMB. *[Courtesy NASA/Goddard Spaceflight Center]*

Plate VII. Galaxy cluster Abell 1689 photographed with the Hubble Space Telescope. *[Courtesy NASA/Hubble Space Telescope Science Institute]*

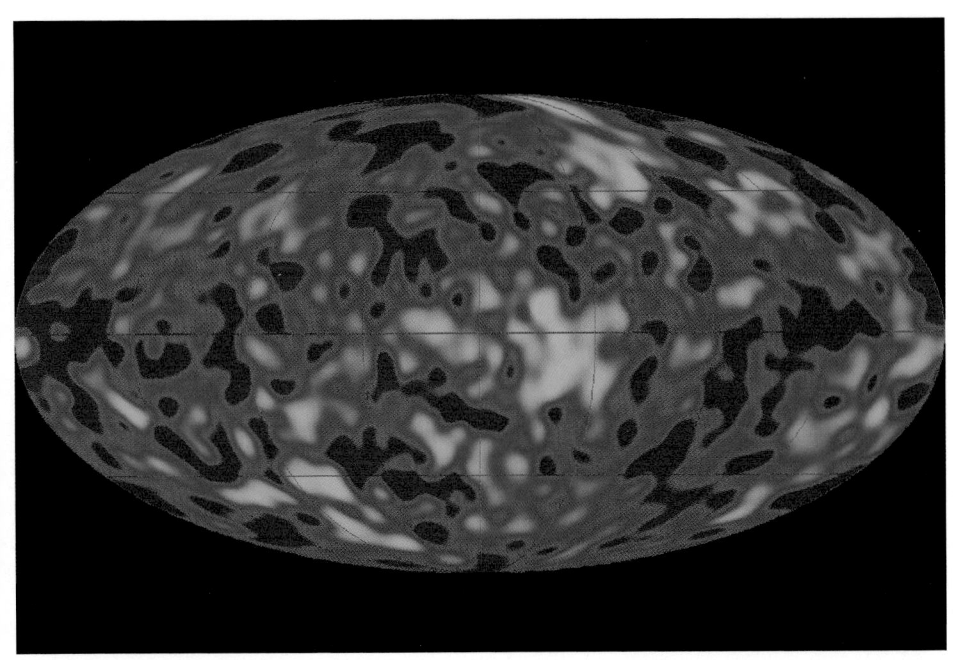

Plate VIII. This COBE image shows the cosmic microwave background fluctuations taken with the Differential Microwave Radiometer. The fluctuations are extremely faint, only one part in 100,000. These density fluctuations are believed to have given rise to the galaxies, clusters of galaxies, and other large structures observed in the universe today. *[Courtesy NASA]*

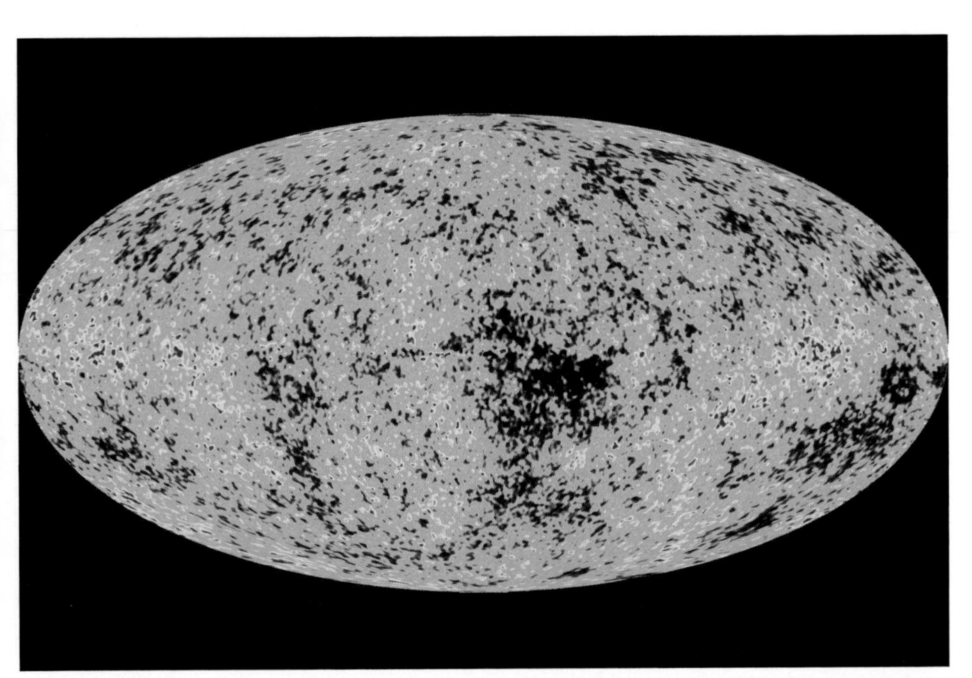

Plate IX. Detailed, all-sky picture of the infant universe using three years of WMAP data. The image reveals the 13.7-billion-year-old temperature fluctuations that correspond to the seeds that grew to become the galaxies. The signal from our galaxy was subtracted. This image shows a temperature range of ± 200 microKelvin. *[Courtesy NASA]*

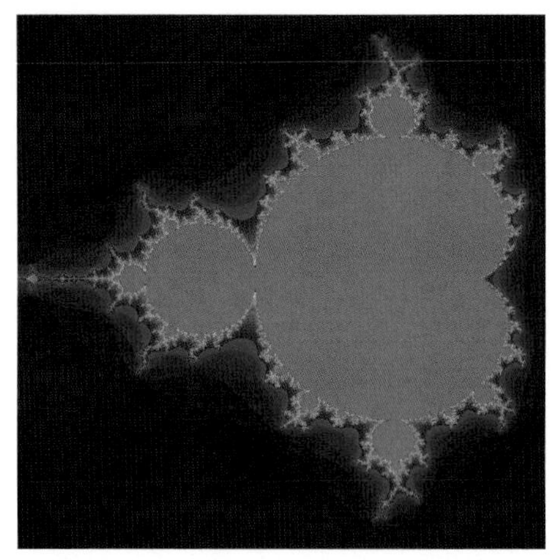

Plate X. According to the eternal inflation model, each stable vacuum configuration in the landscape will generate an ever-increasing number of pocket universes at an ever-increasing rate. The endless generation of pocket universes in a stable vacuum creates a fractal. The fractal shown here is known as the Mandelbrot set, which, like all fractals, exhibits self-similarity at all scales and shows an endless repeated geometry. *[Courtesy of author]*

Plate XII. Resembling a swirling witch's cauldron of glowing vapors, the black hole–powered core of a nearby active galaxy appears in this colorful NASA Hubble Space Telescope image. The galaxy lies 13 million light-years away in the southern constellation Circinus. This galaxy is a Type 2 Seyfert, a class of mostly spiral galaxies that have compact centers and are believed to contain massive black holes. According to the cosmological natural selection model, black holes give birth to new universes. *[Courtesy NASA/Space Telescope Science Institute]*

Plate XI. Sixteen extrasolar planet candidates orbiting several distant stars in our Milky Way galaxy. *[Courtesy NASA/ESA/Hubble Space Telescope Science Institute]*

Plate XIII. A population of infant stars in the Small Magellanic Cloud, some 210,000 light-years away. This nebula contains about 2,500 infant stars that have not yet ignited their hydrogen fuel to sustain nuclear fusion. Populations of young stars such as these offer a unique laboratory for understanding how stars formed in the early universe. *[Courtesy NASA/Space Telescope Science Institute]*

Plate XIV. Hubble Space Telescope and Keck Telescope image of the most distant known galaxy in the universe. Located an estimated 13 billion light-years away, the object is being viewed at a time only 750 million years after the big bang, when the universe was barely 5 percent of its current age. *[Courtesy NASA]*

Plate XV. Laser Interferometer Space Antenna (LISA), NASA and ESA's first dedicated space-based gravitational wave observatory, scheduled to be launched during the second decade of this century. Three spacecraft will orbit the sun in a triangular configuration, separated from each other by 5 million kilometers (3 million miles) to form an interferometer that can detect gravitational waves. *[Courtesy NASA JPL]*

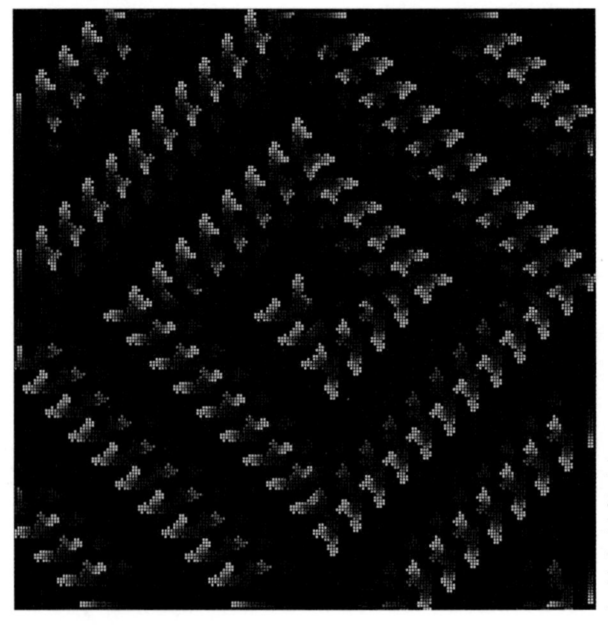

Plate XVI. Cellular automaton generated with a pattern of a few filled cells and the rule that a cell keeps switching colors if it has two neighbors that are also active. *[Courtesy of author]*

greater than 180° (Fig. 6–3). However, if you draw a small triangle on the surface of the sphere, the sides would be nearly straight and the three internal angles would add up to a total that is indistinguishable from 180°. In a small scale, you can apply the simpler rules of Euclidian geometry to study the properties of space. The entire sphere can be represented by the sum of small areas to which the simpler Euclidian geometry applies.

With the compactification of the Calabi-Yau manifold, superstring theory reproduced the mirror asymmetry of the weak interaction and the theory regained its status as a viable theory with all the promise of the unification of the forces of nature. Physicists around the world flocked to join the ranks of string theorists, and it seemed as if it would soon be possible to come up with a consistent theory that could describe the universe and that this description would match the universe that we observe.

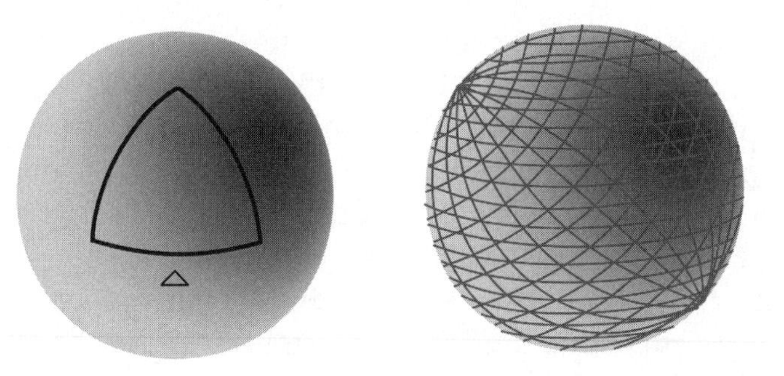

Figure 6-3. The surface of a sphere is an example of a manifold. On a large scale, the sum of the internal angles of a triangle is greater than 180°. However, on a small scale, the surface is nearly flat and follows the rules of Euclidian geometry. In this case, the sum of the angles is equal to 180°. The sphere can be represented by a collection of small flat triangles to which Euclidian geometry applies.

But the euphoria didn't last. Soon, physicists were able to use the Calabi-Yau compactification to generate many possible models that described a universe, each with its own set of laws, its own collections of matter particles, and its own forces. Some could approximate the observed universe, but others were entirely different. Out of the many possibilities, is it still possible to select the one that describes our universe? The answer is no. At this point in the development of string theory, the question of which is the right Calabi-Yau compactification for our universe remains unanswered.

M-THEORY

In addition to the many possible ways to compactify the extra dimensions, physicists realized during the mid-1980s that they could incorporate supersymmetry into string theory in *five* distinct ways, and the result was five different versions of superstring theory. These five versions were called *Type I*, *Type II*, *Type IIB*, *Heterotic-O*, and *Heterotic-B*. Type I was the original version and is the only one that includes open and closed strings.

How can the theory that is supposed to unify all the forces in nature and that has the possibility of being *the* theory of the universe come in five different versions? Either something was wrong with the whole idea of superstrings or the five different versions were actually the same theory in five different formulations. Something similar happened with quantum theory back in the 1920s. In the summer of 1925, Werner Heisenberg developed quantum theory using the mathematical formalism of matrix mechanics. During Christmas of that same year, Erwin Schrödinger developed an alternative version using wave mechanics. In 1927, Paul Dirac developed yet a third and more powerful version. Shortly afterward, Dirac himself demonstrated that the three versions were simply different formulations of the same underlying theory. Could that be the case with the five versions of superstring theory?

As one of the top string theoreticians in the world, Edward Witten was the right person to tackle that question. Born in Baltimore in 1951, he received his doctorate in physics from Princeton University. After a postdoctoral fellowship at Harvard, he became a junior fellow at Harvard. In 1982, he was awarded a MacArthur Fellowship. He returned to his alma mater in 1980 and later moved to the Institute for Advanced Studies in 1987, where he is a professor of physics. In 1990, he won the Fields Medal—considered to be the mathematics Nobel—for his work in superstring theory.

In 1995, Witten showed that, indeed, the five different superstring theories were actually five different mathematical descriptions of the same single theory and proceeded to develop a more comprehensive version that would encompass all the other versions. He called his new approach *M-theory*. To accomplish his synthesis, Witten needed to add another spatial dimension. M-theory is described in *eleven* dimensions, ten of space and one of time, and, as it turned out, these eleven dimensions should have been included in the earlier versions all along, but theorists hadn't noticed the need for all of them in their equations.

After Witten's development of M-theory, Joe Polchinski of the University of California at Santa Barbara showed that M-theory includes not just two-dimensional strings, but higher dimensional objects. As an open string vibrates, the two ends move, describing a two-dimensional surface or membrane (Fig. 6–4). Polchinski's calculations showed that these two-dimensional membranes, or *branes*, as he called them, could not only be described from within M-theory but were actually required to exist by the mathematical formalism of the theory. What was more, M-theory also predicted that higher dimensional branes—*p*-branes—with up to ten dimensions should also exist. When *p* is one, M-theory describes a string; when *p* is two, three, or ten, M-theory describes a two-, three-, or ten-dimensional membrane. We can only visualize a string and a two-dimensional membrane. Higher dimensional sheets can only be described mathematically. The three dimensions that we are able to experience directly in our universe reside on a three-brane.

Branes, like strings, have physical reality and their own physical properties. One particularly interesting property of branes is that they can hold electric charges such as those that give rise to fields or the charged Yang-Mills force carriers. In M-theory, photons are the vibrations of open strings and reside on three-branes and are therefore confined to our three-dimensional space. If photons move in this three-brane, the brane is transparent to photons and as a consequence are invisible to our eyes and to any instrument operating in any region of the electromagnetic spectrum. No optical telescopes or microscopes; no radio, x-ray, microwave, infrared, or ultraviolet telescopes could detect it. Our three-brane is also undetectable by any means that make use of any of the particles of the Standard Model, matter or mass particles. The reason is that the matter particles, the quarks and the leptons, are also vibrations of open strings and reside on the three-brane. Such is the case also with the force particles, such as the gluon that carries the strong interaction or the *W* or *Z* bosons, the carriers of the weak interaction. All these particles reside on the three-brane, which is then transparent to them, making them unusable as detectors of the three-brane.

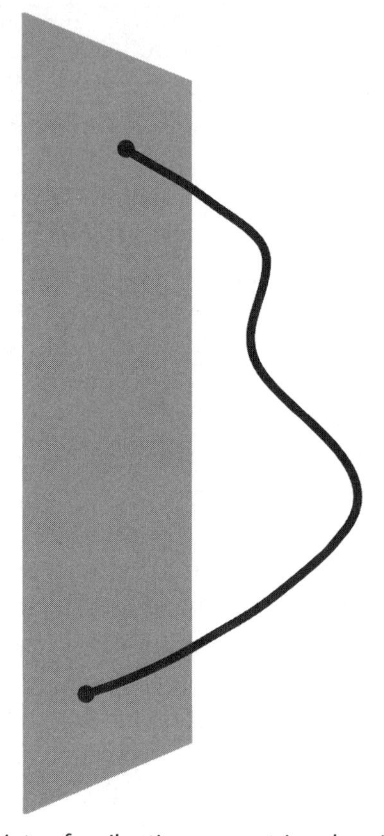

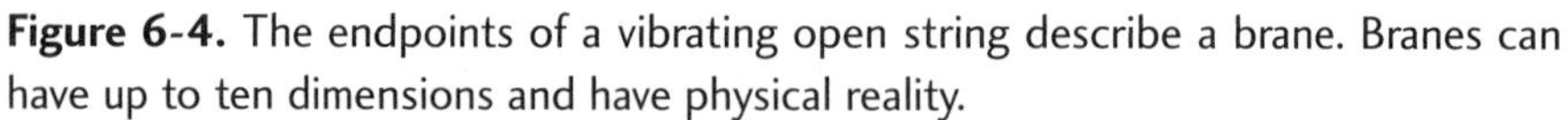

Figure 6-4. The endpoints of a vibrating open string describe a brane. Branes can have up to ten dimensions and have physical reality.

Will the three-brane remain always out of our reach? According to M-theory, the graviton is the lowest mode of vibration of a closed string. Since closed strings have no ends, they can't be attached to branes and must exist on the full ten-dimensional space. Gravity then acts throughout all of space and the graviton may be used as the probe to detect the higher dimensions of space.

The M-theory view of the universe is that the fundamental constituents of matter are the multidimensional branes. The superstring is nothing more than a brane with all the dimensions beyond one being rolled up in a much smaller scale. What the early string theorists saw as superstrings were these curled up branes seen through the blurry lens of their more primitive and incomplete theories. M-theory sharpens the focus and increases the magnification, allowing us to discover that the building blocks of matter are branes.

The Work Is Not Done

The description of the laws of physics started in the previous chapter is now complete. We have gone from the well-understood depiction of the matter and force particles of the Standard Model, guided by the rules of quantum mechanics, to the frontiers of knowledge, where we encountered a master theory that promises to unify all the forces of nature and explain how the universe is put together.

The Standard Model, the most comprehensive theory ever developed, is a description of the structure of the universe whose theoretical predictions have been proven to a high degree of accuracy. Using the powerful concepts of symmetry, the theory unifies the electromagnetic and weak nuclear interactions into one single electroweak force. The theory showed how the Higgs mechanism can be used to generate the masses of the mediators of the weak force and showed us that the different masses of the other elementary particles found in nature are not arbitrary values with unknown origin but actually came naturally and automatically from the Higgs mechanism.

The success of the electroweak unification whetted physicists' appetites for a grand unification of all the nongravitational forces. And the extended supersymmetric Standard Model has brought them closer to achieving that goal. But the complete unification of *all* the forces of nature is beyond the capabilities of the Standard Model, extended or not. The task of achieving that ultimate goal of physics lies now with M-theory.

Much remains to be done. We don't know why there are three generations of matter and force particles in nature and why everything that we know about the universe can be described by the first generation. The supersymmetric Standard Model has not yet fully achieved the grand unification. The discovery of the Higgs particle awaits the experiments of the Large Hadron Collider (LHC) at CERN. The mass of the lightest Higgs boson should be within reach of LHC. Similarly, the masses of the superpartners should be in the range of this collider as well. The discovery of these particles would tell us if the early universe was supersymmetric. And no single M-theory prediction has ever been proven. But no prediction has been refuted, either. The energies needed to validate the predictions are beyond our current technology.[9] However, as Ed Witten said of string theory, it "has the remarkable property of predicting gravity."[10] And that's why string theory and the more general M-theory need to be considered very seriously in spite of the present lack of experimental proof. We don't know if M-theory is correct, but it is the first theory ever to bring together gravity and the three Standard Model forces. When technology advances make it possible to test the theory, we will know whether it was the correct approach.

The extraordinary success of the Standard Model tells us that it is undoubtedly correct. We know it isn't complete, but we know that what it does is accurate. If the lightest superpartners are discovered with the LHC, we would know soon whether supersymmetry is correct and that would tell us if the supersymmetric Standard Model is viable. The grand unification of the nongravitational forces will soon follow. However, we know even now that the extended Standard Model cannot incorporate gravity. For that we must follow the route of M-theory. Although it hasn't been done yet, it should be possible one day to automatically derive all the properties of the Standard Model from M-theory. But before we declare victory, some of the theory's predictions must be validated experimentally.

ARE THE LAWS OF PHYSICS FINE-TUNED FOR LIFE?

My description of the laws of physics is complete, but at the present time our understanding of those laws is anything but complete. We have achieved enormous successes but we're far from understanding all the details of what makes the universe tick. We don't know exactly how the laws of physics—the watchmaker of the universe—operate. What we do know is that we have caught a glimpse at how they work.

Experimental evidence tells us that the details that we do know are accurate and that we're making great inroads toward achieving a complete understanding of the way the watchmaker of the universe operates. At times, we will reach dead ends, but we know that we will turn around to the previous fork in the path and blaze a new path that takes us closer.

One thing physicists agree on is that the laws of physics are interconnected with extreme precision. Even a minute deviation in a single fundamental parameter spoils the entire fragile equilibrium. In 1953, the British astrophysicist Fred Hoyle pointed out that we owe our own existence to a particular high-energy configuration of the nucleus of carbon-12. If that particular quantum state of carbon didn't exist, the stellar nucleosynthesis mechanism would not have produced any carbon and the universe would be composed of hydrogen, helium, lithium, and beryllium only. That would make a very different universe from the one we inhabit. The main difference would be that carbon-based life—the only one we know has led to advanced organisms—would not be possible in a universe in which the laws of physics did not allow for that specific energy configuration.

Recall from chapter 3 that George Gamow and his collaborators had reconstructed the formation of the light elements from hydrogen to helium during the nucleosynthesis phase of the big bang. When it was discovered that element formation during the early universe had stopped with helium, the obvious question was, how did the heavier elements form? Arthur Eddington gave us the correct answer when he proposed the idea that the "stars are the crucibles in which the lighter atoms which abound in the nebulas are compounded into more complex elements."[11] But it was Hoyle who actually made the calculations of the different physical conditions during the life of stars and developed the model of how stellar nucleosynthesis might occur for nuclei beyond helium. But he also reached a stumbling block. There was no clear mechanism for producing carbon by combining lighter nuclei. The first method that came to mind was the fusion of three helium-4 nuclei, with two protons and two neutrons each, to form the nucleus of carbon-12 with six protons and six neutrons. However, calculations showed that the likelihood of these three particles colliding simultaneously was deemed to be almost zero. What if only two helium nuclei collided first to form beryllium-8, followed by a collision with a third helium nucleus? That mechanism (illustrated in Fig. 6–5), first proposed by Edwin Salpeter of Cornell University, was not seen as being impossible, but, unfortunately, beryllium-8 is unstable and decays with a half-life of less than 10^{-16} second. Closer examination showed that the energies of the colliding nuclei didn't match—there was no *resonance* at that energy. The energy of helium-4 and beryllium-8 together is greater than that of carbon-12 and there was no mechanism to get rid of the extra energy that this excess mass created (from $E = mc^2$).

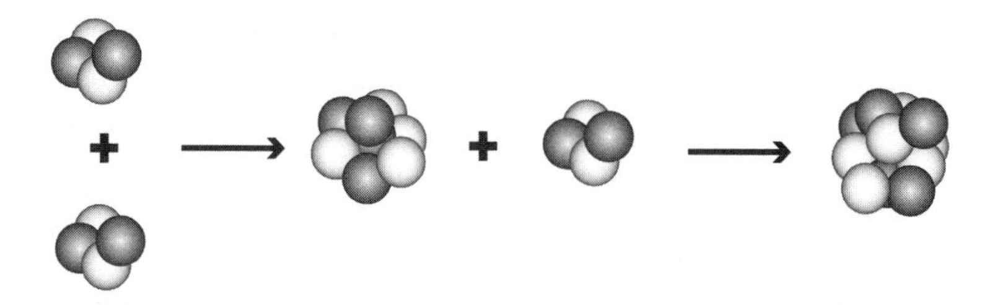

Figure 6-5. The mechanism for carbon-12 production in the interior of stars: Two helium-4 nuclei combine to form beryllium-8. A third helium-4 nucleus combines with the beryllium-8 nucleus to produce carbon-12. However, there was an energy imbalance that needed to be resolved.

However, quantum mechanics tells us that nucleons in a nucleus exist in quantized energy states (similarly to the energy states of electrons in an atom). Normally, the nucleus is in its ground state, with the lowest energy configuration. But the nucleus can be excited to a higher-energy state. Nuclear physicists study these energy states of nuclei in an effort to understand the different interactions that take place between nuclei. Fred Hoyle had calculated that the temperature in the interior of a large star should be about 100 million degrees[12] and with that, he was able to work out the value of the kinetic energy of the colliding nuclei. Armed with that information, Hoyle predicted the energy of the exited state of carbon-12 that would exactly match the total energy of the helium and beryllium nuclei combined.[13] He then approached the nuclear physicists who told him that such a state had never been observed. At his insistence, William Fowler at Caltech reluctantly agreed to perform additional experiments to look specifically for this very specific exited state. After a few days, Fowler was able to find the missing state with exactly the energy that Hoyle had calculated. Even today, this excited state of carbon-12 is called the "Hoyle resonance," a small consolation prize for being left out by the selection committee for the 1983 Nobel Prize in physics, awarded only to Fowler for "his work on the formation of the chemical elements."

However, the discovery of this excited state in carbon-12 did not by itself assure our existence. Most of the time, the excited state of carbon-12 decays back into its original constituents, beryllium-8 and helium-4. In fact, it does so all but 4 times out of 10,000, and only these 4 times produce carbon-12 in its ground state. This is the configuration needed to form complex organic molecules that eventually end up in the DNA that makes life possible on Earth.

The existence of this excited state in carbon-12, the instability, and its rate of decay are controlled by the electrostatic and strong nuclear forces. This excited state is actually controlled by the tug-of-war between the electrical repulsion of the protons inside a nucleus and the attraction of those protons due to the strong nuclear force. The strength of the electrical repulsion is controlled by a fundamental constant of nature called the *fine structure constant*. This constant is defined as the combination of the charge of the electron, the speed of light in a vacuum, and Planck's constant, and it has no units. If the value of the fine structure constant or the strength of the nuclear force had been slightly different, the formation of carbon-12 would change dramatically. A change of only a few percent in these values and there would have been no organic chemistry, and hence no life.

It seems as if the laws of physics teeter in a precarious equilibrium that is precisely what is needed for life to exist. Are these laws fine-tuned to assure our existence or is it all a big coincidence? If you think that this could be a coincidence, hang on, because we're going to encounter the biggest coincidence of them all, Einstein's cosmological constant.

The Cosmological Constant

In 1917, Einstein derived a model of the universe from the equations of general relativity. Recall that when his original model resulted in a dynamic universe, Einstein decided to add his cosmological constant to the equations to provide a static solution to his model. By adding this flexibility to his model, Einstein was able to describe the astronomical observations obtained at the time. However, when Hubble discovered that the universe was expanding, Einstein regretted having spoiled his equations with the additional term, calling it the greatest blunder of his life.

Recent astronomical observations of the rate of expansion of the universe show that the cosmological constant was no blunder. As Steven Weinberg has said, "There is no reason *not* to include a cosmological constant in the Einstein field equations."[14] Einstein's relativity postulate states that the laws of physics are independent of the frame of reference used in describing the laws. The relativity postulate is a symmetry principle. And, as Weinberg points out, the only term that can be added to Einstein's field equations of general relativity without violating this symmetry is precisely the cosmological constant.

But what is the cosmological constant? According to Einstein's general relativity, the cosmological constant is the same as antigravity. Let's see how that works. In Newtonian physics, masses create gravity. The sun, the Earth, all the planets, galaxies, stones, and apples attract each other with a force that depends on the masses of the interacting objects and their separation distances. This is positive gravity. In Einstein's physics, $E = mc^2$ tells us that energy also creates positive gravity, since energy can be converted into mass. But Einstein's universe goes further with gravity. According to general relativity, pressure also generates gravity. The normal, run-of-the-mill pressure that pushes outward—like the pressure in your bicycle tires or the pressure of the air—generates positive gravity. What would happen if pressure were negative? In general relativity, negative pressure creates negative or repulsive gravity: in short, antigravity!

When Einstein added his cosmological term, he realized that his equations allowed for negative gravity and negative or repulsive gravity was just what he needed to counterbalance the positive attractive gravity of all the masses in the universe that were making his model unstable. And that's what he added to his equation. His cosmological constant is repulsive gravity, or antigravity.

The cosmological constant explains the recent discovery of the variable rate of expansion of the universe. In 1998, Saul Perlmutter of the

Lawrence Berkeley National Laboratory, Brian Schmidt of the Australian National University, and their research teams discovered, through careful observations of supernova explosions, that for the past 5 billion years the expansion of the universe has accelerated. In 2002, Adam Riess of NASA's Space Telescope Science Institute used the newly installed Advanced Camera for Surveys on the Hubble Space Telescope to confirm that for the first 9 billion years, the expansion of the universe was slowing down and that after that time, the slowdown stopped and the universe began to expand at an ever-increasing rate. What caused this seemingly strange change in the universe's expansion? The explanation lies with the cosmological constant, with negative gravity.

The rate of expansion of the universe depends entirely on gravity. Positive, attractive gravity caused by the masses of all the galaxies and by the energy in the universe slows down the expansion. As the universe pulls apart, this gravitational attraction decreases, since it depends on the distance between the galaxies. But the negative, repulsive gravity of the cosmological constant doesn't depend on distance and remains constant during all phases of the expansion. When the universe was young, attractive gravity was stronger than the cosmological constant repulsion. As the universe expanded, the attractive gravity decreased and at one point, 9 billion years after the big bang, it matched the strength of repulsive gravity. But the universe was still expanding at the time, and inertia kept the galaxies moving away from each other. After a while, repulsive gravity became dominant and the expansion of the universe began to accelerate.

Einstein's blunder turned out to be the only feasible explanation to the variation in the expansion of the universe. But there is more. The cosmological constant is the same as the energy of the vacuum (which today is also called *dark energy*). Recall that the vacuum is filled with virtual particles that contribute energy to the vacuum. Although all virtual fermions, like electrons, neutrinos, and quarks, contribute negative energy whereas virtual bosons, like the photons, gluons, and Higgs particles, contribute positive energy, the total sum of all these contributions isn't zero. The energy contribution from each one of the particles is different and if we were to add up all the energies for all the virtual fermions and all the virtual bosons in the universe, negative values for the fermions and positive values for the bosons, what we come out with is not only not close to zero, but huge. The only way for these energies to come out to zero is if the world were supersymmetric. But even if the supersymmetric partners existed in the past, we don't think that they exist today. The universe today isn't supersymmetric.

The experimental discovery of the speeding up of the expansion of the universe 9 billion years after the big bang requires that all the vacuum energy contributions cancel out to 120 decimal places. Cancellation to that accuracy is impossible to comprehend. If we take the total number of sand grains in all the beaches of the world and charge half of them with positive charge and the other half with negative charge so that charge cancels out, producing a neutral pair, this cancellation is only to some 24 decimal places. What if we take the total number of atoms in the world or, better yet, the total number of elementary particles and performed a similar cancellation? That would be to 51 decimal places. Even if we consider the estimated total number of elementary particles in the observable universe, we would get "only" 80 decimal places. A cancellation to 120 decimal places requires extremely delicate fine-tuning!

IS GOD IN THE DETAILS?

It appears as if the existence of life on Earth depends on a fragile equilibrium of the forces and constants of nature. The excited state of carbon-12—the first example of this equilibrium—is one of many examples. The unlikely connection between the million-year trek of a photon in the interior of the sun and the human eye that I described in chapter 1 is another one. Unlikely as these two examples may seem, they both pale in comparison with the improbable balancing act of the tiny value of the cosmological constant. As Leonard Susskind has pointed out, a physicist rather accepts that there is some deep, hidden principle in nature that we haven't yet discovered and that would make the cosmological constant *exactly* equal to zero.[15] But the perfect cancellation of 119 decimal places so that, beginning with the 120th place, nonzero values would appear is very hard to understand.

Is this delicate balance of the laws of physics that make the universe suitable for life evidence that the universe was tinkered with so that it would lead to our existence? In the absence of a complete theory capable of explaining the exact value of the cosmological constant and the fragile equilibrium of the laws of physics, it would appear as if the laws of physics, the watchmaker of the universe, would require their own watchmaker.

Is this really the case? Have we found a job that only a supernatural designer can handle? Is the fine-tuning of the cosmological constant beyond the reach of science? In the following chapters, we will explore possible solutions to the cosmological constant problem, a problem that may be the biggest problem in physics today. Some of these exciting new ideas make use of the innumerable models of the universe that can be derived from M-theory to construct a *multiverse*, a metauniverse that contains all the possible universes that the theory is capable of describing. If these ideas are correct, we don't have to arrive at an automatic and unique description of our own laws of physics and of the cosmological constant, since all allowed descriptions exist.

Chapter 7

A LANDSCAPE OF POCKET UNIVERSES

FRAGILE EQUILIBRIUM

The laws of physics together with extremely accurate experimental observations have shown us that the universe is precariously balanced on an edge so sharp that you'd need 120 decimal places to describe its deviation from a perfect edge. Any departure from this value even by one order of magnitude would have prevented the formation of galaxies during the early universe. No galaxies means, of course, no stars and no planets, and therefore, no life. The delicate balance of the laws of physics and of the constants of nature seems to indicate that the universe was designed for life. But was it? Let's look deeper into this problem.

Is life as we know it the only possible life? Life on Earth, the only example we know, is based on carbon-rich (organic) compounds and water. The element carbon, unlike any other known element, possesses an extraordinary ability to combine with other atoms. No other element comes close to carbon in the variety of the compounds that it can form and in their complexity. Liquid water provides a stable medium for these complex compounds to interact and form even more complex systems. But are we being too parochial? Does life need to be carbon and water based? Could life on other worlds be based on silicon, for example, which shares similar chemical properties with carbon (it belongs to the same group in the periodic table of the elements)? Silicon doesn't really match carbon when it comes to forming compounds. Carbon is abundant in the universe because it is a by-product of stellar evolution. Silicon, on the other hand, was produced at a later stage, during the carbon-burning phase in the life of high-mass stars; along with phosphorous and magnesium, silicon is a by-product of the oxygen-burning process and is not as abundant as carbon.

On the other hand, there is a water substitute for the silicon world: hydrogen fluoride.[1] Like water, hydrogen fluoride can dissolve other molecules and could be a medium for the formation of more complex systems. But fluoride is not common throughout the universe. Organic compounds are. They've been found in the interstellar gas and dust, in meteorites, and in many of the large planets of the solar system.

Our own carbon-based life appeared on Earth as soon as the conditions were right. It seems as if the origin of life was an inevitable consequence of chemical processes. However, the emergence of life from organic matter was not an easy process, even with the compound production ability of carbon and the presence of water. The laws of physics in our universe are such that carbon-based life seems to be the most viable possibility.

The equilibrium of the laws of physics in our universe appears to be fragile. If the universe had evolved with a different set of laws, one could argue that these laws could also give rise to forms of life based on elements other than carbon that could possess comparable or perhaps even better properties of combination with certain other elements, thus leading to complexity and eventually to some form of life. In a universe with a different fine structure constant and a different strength of the nuclear force, the table of the elements would be different, if there were elements at all. At present, we don't have a good way to know if any other combinations of the laws of physics could lead to intelligent life.

MULTIPLE UNIVERSES

Fragile or not, the laws of physics are described by M-theory—the powerful theoretical construct that unifies all the forces of nature, including gravity. M-theory, in principle, encompasses the Standard Model, the extraordinarily accurate description of the microscopic structure of the universe whose predictions have been corroborated experimentally to a high degree of precision. As described in chapter 6, M-theory promises to explain the workings of the world, from the subatomic to the macroscopic scale.

However, M-theory doesn't exactly describe our universe. It contains the descriptions of innumerable universes. Estimates of the possible number of universes that the theory is able to describe reach the unimaginable number of 10^{500}. This number is so large that we don't have an easy way to compare it to anything. For example, the number of seconds that have elapsed since the big bang is "only" about 10^{17}. One possible universe for every second that our universe has existed takes us only to a minute fraction of the possible universes described by M-theory. We want big numbers. What if we take the number of atoms that make up the Earth? That's still only about 10^{50}. And the number of atoms in the entire Milky Way galaxy? That's just about 10^{68}. Not even the number of elementary particles in the entire visible universe gets us close, since that's about 10^{80}. The theory describes six possible universes for every atom in our own universe!

How did this multiplicity of universes that strain the imagination come about? Recall that in M-theory, the fundamental components of matter are the multidimensional branes that exist in eleven dimensions, ten of space and one of time. Dimensions over four—our familiar spatial dimension and time—are rolled up in a tiny circle. The circumferences of the circles of the extra dimensions determine the strength of the different forces and constants. Recall also that in M-theory, the compactification of the unseen dimensions is described by Calabi-Yau shapes. These six-dimensional shapes assure us that the mathematical properties required by M-theory are satisfied. Each Calabi-Yau shape, with its own set of laws and constants, describes a different universe.

Calabi-Yau shapes can have holes. A torus (a doughnut-shaped figure), for example, has one hole. A torus twisted into a figure eight has two holes, and a pretzel-shaped figure has three. Calabi-Yau shapes can vary in the number of holes, the length and thickness of the loop, and the angle of the different loops relative to each other (Fig. 7–1). In addition to these geometrical parameters, the variation in the field lines from the different forces that can flow through the holes provides another parameter. These field lines are a powerful concept invented by the nineteenth-century physicist Michael Faraday to visualize the strength of the electric and magnetic forces at every point in the vicinity of an electric charge or a magnet. The common school demonstration with iron filings sprinkled on a piece of paper laid over a magnet is probably the best illustration of magnetic field lines. Field lines emanating from the north pole of a magnet, for example, circle back toward the magnet's south pole. If a torus is brought to the vicinity of this magnet, the number of field lines that cross through the hole depends on the torus's proximity to the magnet. The number of field lines crossing through the loop is called the *magnetic flux*. Electric and magnetic fluxes are described mathematically in three-dimensional space by Maxwell's equations. The mathematical description of fluxes through the six-dimensional spaces is much more complicated, but you can still think of them as field lines threading through the doughnut holes of these Calabi-Yau shapes.

Figure 7-1. A torus or doughnut is a simple geometrical shape with one hole. A figure 8 has two holes and a pretzel has three. More complex six-dimensional geometrical shapes called Calabi-Yau manifolds can also contain holes. The number of holes, together with the thickness and the size of the loops, play important roles in the compactification of the unseen dimensions of the universe.

The curvature of the compacted dimensions, the number of holes, the length of the loops, and the number of flux lines flowing through these holes form a specific configuration of spacetime with a certain potential energy. When the configuration has no matter in it, the potential energy is the energy of the vacuum. If we consider only one the parameters—the diameter of the loop in each Calabi-Yau shape, for example—the vacuum energy would have hills and valleys as the energy varies as a function of that parameter (Fig. 7–2). This vacuum energy will tend to reach the minimum values, just like a ball rolling off a hill with similar hills and valleys will settle at the bottom of one of the valleys. The valley in which it finally settles depends on its initial conditions. If the ball is rolling from a high point on the left slope, it will roll off through all the valleys and peaks and end up rolling all the way to infinity. If it rolls off from the second hill, it will settle on the third valley, since it wouldn't have enough energy to climb the last peak.

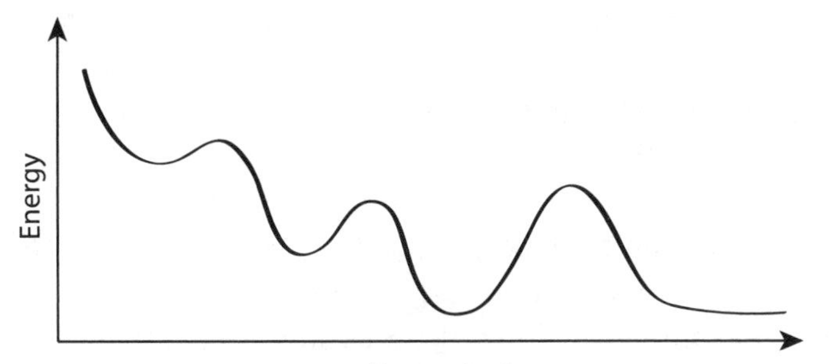

Figure 7-2. The energy curve showing the possible values of the energy of the vacuum as a function of one single parameter, the diameter of the loop in this case.

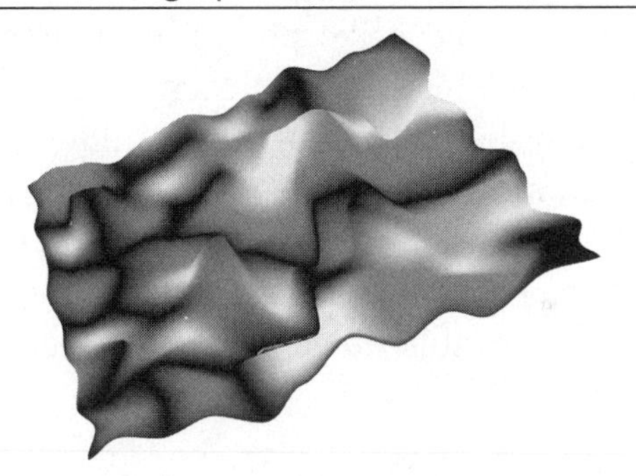

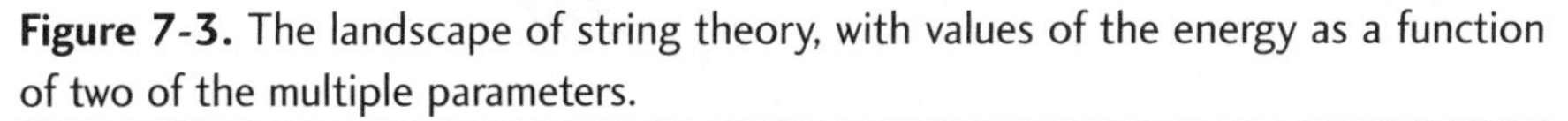

Figure 7-3. The landscape of string theory, with values of the energy as a function of two of the multiple parameters.

Since there are many parameters, the vacuum energy curve is actually a multidimensional terrain with mountains, slopes, and valleys; the *landscape* of string theory, as Leonard Susskind of Stanford University calls it (Fig. 7-3).[2] The valleys in the landscape are the stable vacuum configurations of spacetime, where a set of physics laws and constants of nature give rise to a pocket universe.

How many stable configurations of the vacuum exist? In 2000, Raphael Bousso, then a postdoc at Stanford University, and Joseph Polchinski of the University of California at Santa Barbara, calculated the possible number of solutions provided by the different combination of parameters in the six-dimensional Calabi-Yau shapes.[3] Based on calculations by other researchers of the number of possible holes that give an upper limit of five hundred, Bousso and Polchinski determined that the number of flux lines wrapping around each hole cannot exceed ten, since, according to their calculations, a larger number would make spacetime unstable. With those values, there would be 10^{500} possible stable vacuum configurations where a universe could exist.

Each valley in the landscape establishes a unique set of the laws of physics, particle masses, and constants of nature, as well as its own cosmological constant. Since physicists have shown that the Calabi-Yau shape can reproduce the mirror symmetry violation observed in our world, they are hoping that the Standard Model that describes our universe could one day be derived from M-theory.

But is our valley the only possible valley where a universe can exist in the landscape? Before Susskind's proposal for the existence of this vast and complex landscape, physicists thought that the multiplicity of models allowed by M-theory, each describing a possible universe, would one day be shown to be just different versions of the same unique model, much like Witten had done in 1995 with the five early versions of superstring theory. Today, that possibility doesn't appear to be viable. For this reason, Susskind's proposal that each vacuum represents a possible universe has received some attention.

DO OTHER UNIVERSES EXIST?

With 10^{500} possible stable vacuum configurations, is our universe the only one in existence or are these configurations filled with other universes? Can we even find the answer to this question? One possibility would be that our valley in the landscape is unique and that there is a mathematical principle embedded in the underlying laws of the universe that steered our universe toward that special place. However, we may never know what the underlying laws are, since they would govern the entire landscape and reach beyond our own universe. Discovering that hypothetical mathematical principle may prove to be impossible. Even if this principle existed, it would require the extreme fine-tuning of selecting one valley among 10^{500} with the precise small value of the cosmological constant that allows life to appear, survive, and evolve.

Bousso and Polchinski, as well as Susskind, think that they have a better idea. According to them, all valleys in the landscape are populated by pocket universes and ours is simply one of them. In their 2000 paper, Bousso and Polchinski proposed a mechanism for this populated landscape, or *multiverse*, as they call it.[4] In their view, the application of quantum mechanics and general relativity to the multiple solutions of M-theory showed that the configuration of compacted dimensions can jump from one valley to another and that a rapid growth of spacetime allows these configurations to exist simultaneously as independent pocket universes.

Remember that each vacuum configuration in the landscape has its own cosmological constant. However, according to quantum mechanics, the vacuum configurations are metastable and can decay at any moment. Metastability is a well-known phenomenon in physics. As explained in chapter 4 in connection with the inflationary theory of the universe, water can exist in a metastable state when it is supercooled, as is the case with freezing rain. Under certain circumstances, for example, when the temperature of the water is lowered at a very fast rate, water can remain liquid at 20 degrees below its freezing point. This supercooled state of water is unstable and a disturbance will trigger the formation of ice. Adding a small chunk of ice to supercooled water will start the process of nucleation and the water will start to form crystals around the added ice. Supercooled water will also freeze by itself after some time, usually long, when some of the water molecules moving at random suddenly organize themselves in a crystal configuration. If the crystal's mass is larger than some critical mass, it will trigger the nucleation process. Because the crystal grows at a relatively rapid pace, the spontaneous ice formation in supercooled water is called bubble nucleation.

Just as supercooled water can spontaneously freeze by forming expanding bubbles of ice, the vacuum configurations in the landscape can decay by forming bubbles with an energy density slightly lower than the energy density of the originating vacuum. The decay is caused by a spontaneous change in the configuration of the Calabi-Yau shape that describes the compactification of the unseen dimensions in this particular valley. If the number of flux lines is reduced by one, for example, the energy density is reduced by a relatively small amount (Fig. 7–4). If the new lower value of the energy in the bubble that forms matches the energy of a nearby valley, quantum fluctuations allow the bubble to tunnel through the surrounding mountain ridge to that nearby valley.

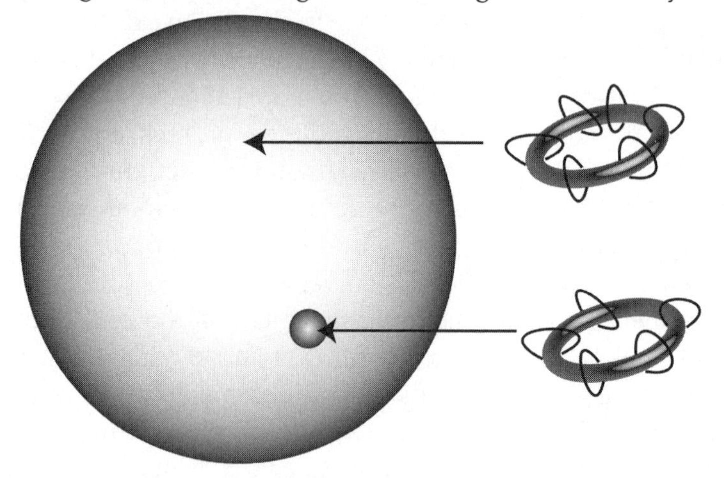

Figure 7-4. Random fluctuations in a stable vacuum, represented here by the large sphere, allow a small region to form (the small sphere). The original vacuum has a manifold of the curled-up extra dimensions, represented here by a torus with six flux lines wrapped around. When one of the flux lines decays due to quantum fluctuations, the small region with slightly lower energy develops. The lower energy of the bubble corresponds to the energy density of a nearby vacuum and the bubble tunnels through the hills surrounding the old vacuum to the nearby vacuum that matches its energy density.

But how does this quantum tunneling through the mountain ridge work? The landscape is a topographic map of the energy density of the vacuum as a function of the different parameters that determine the geometry of the hidden dimensions, so that in this map, higher altitude represents higher energy. The stable vacuum configurations (the dips in Figs. 7–2 and 7–3) are regions of lower-energy density than the surrounding hills. The metastability of the vacuum configurations comes from the ever-present vacuum fluctuations. These fluctuations are controlled by the Heisenberg uncertainty principle so that a given region in a particular valley can spontaneously experience a sharp increase in energy that lasts for a very short time (Fig. 7–5). The uncertainty principle sets a limit to both quantities such that the product of the energy increase and the time this energy increase lasts must be at the most equal to Planck's constant divided by 2π. For a very short time, then, the energy can take very large values. When these values reach levels larger than the higher-energy peaks surrounding the valley, the mountain ridge can be breached, a phenomenon known as *quantum tunneling*. Over time, random quantum fluctuations can allow the bubble to jump over the hill that surrounds the valley. When that happens, the bubble with a lower-energy configuration begins to expand rapidly, growing to a volume billions of light-years in diameter. Because of this rapid expansion, there is a danger that the bubble soon overtakes everything. However, space itself is expanding, and it does so at a much faster rate.

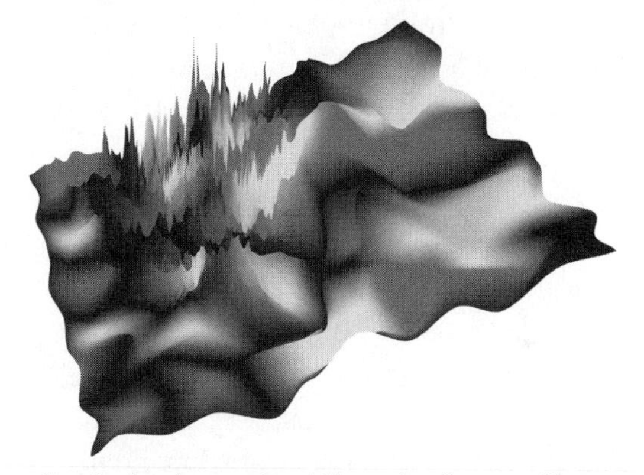

Figure 7-5. Normally, the landscape has hills and valleys, as illustrated in Fig. 7-4. In some locations, quantum fluctuations momentarily produce very sharp and tall peaks that tower over the highest mountains. The formation of those tall peaks is controlled by the Heisenberg uncertainty principle. The high-energy peaks allow the tunneling of the bubble to nearby valleys.

Supercooling, bubble nucleation, rapid expansion of space—all this should sound familiar to you. These are the concepts behind the inflationary theory described in chapter 4, except that these ideas were used to explain the decay of the false vacuum, the metastable state that gave rise to our own universe, not the decay of the different vacuum configurations in the landscape. However, in 1987, Andrei Linde of Stanford University expanded the inflationary theory of the evolution of the early universe to explain the evolution of the universe beyond the visible.[5] This model—developed further by Linde, Alexander Vilenkin of Tufts University, and by Alan Guth of the Massachusetts Institute of Technology—now known as *eternal inflation*, is very similar to the multiverse model of Bousso and Polchinski.

In the eternal inflation model—or the landscape model—the decay of the vacuum is exponential, like the decay of a radioactive nucleus. The half-life of vacuum decay is estimated to be around 10^{-35} to 10^{-30} second. At the end of one half-life, on average, half of the region of a stable vacuum valley in the landscape would have decayed to nearby regions of lower-energy density, with the other half remaining in the original vacuum configuration. At the end of the second half-life, one quarter of the original region would remain and the rest would have decayed.

In the inflationary model, the region in the stable vacuum that hadn't decayed would continue to expand exponentially. But the rate of expansion of the stable vacuum region would be greater than the rate at which it would decay so that after one half-life, the volume of the stable vacuum that would remain would grow to a volume much larger than the original volume. Inflation would enlarge the decaying stable vacuum so that it would continuously grow. The bubbles that form at each decay would also increase in size, and both the old and the new regions would grow forever. To illustrate the battle between these two expansions, imagine a species of free-floating algae with a very fast reproducing rate. One genus of microscopic green alga called *Hydrodictyon*, for example, forms a meshlike web that floats on the surface of ponds. *Hydrodictyon* doesn't simply grow undisturbed, because mayflies, midges, stoneflies, and other aquatic insects feed on it. However, for our analogy, imagine that we remove all of these predators from a lake where this alga is growing and allow it to reproduce freely. Soon, we'll have growing islands of this alga all over the surface of our lake. In time, these islands will run into each other and the surface of the lake would end up completely covered with the alga. To prevent that, we continuously add fresh water to the lake, allowing the lake to expand with the new water. If we add water at a faster rate than the rate at which the alga reproduces, the growing islands of alga will never run into each other, even though they are continuously growing. In fact, any one point on the surface of the water will eventually be overtaken by the alga, but the alga islands will never connect with each other.

This analogy describes the eternal inflation scenario. The growing alga in the freshwater lake are the bubbles that spontaneously form in the stable vacuum configurations of the landscape and tunnel to the nearby valleys. If we now represent a stable vacuum configuration of a given energy—a valley in the landscape—as a small sphere, as time progresses, the region inflates and decays, giving rise to bubbles of stable vacuum configurations that expand into what Guth calls *pocket universes* (Fig. 7-6). These pocket universes grow to become large universes, undergoing an initial inflationary period followed by a big bang expansion. Our observable universe is thought to be a small fraction of a pocket universe.

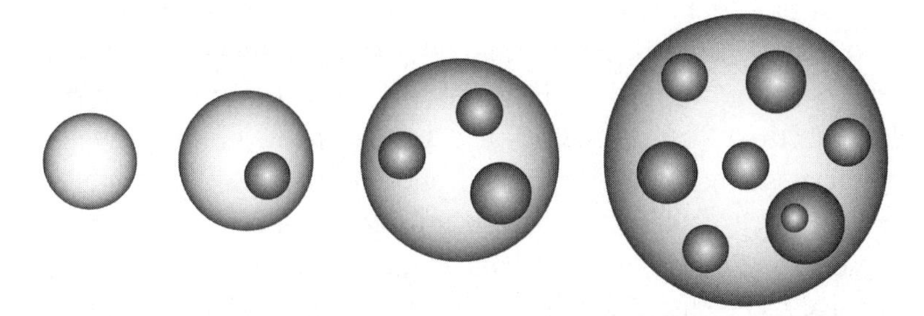

Figure 7-6. A stable vacuum configuration in the landscape (left) gives rise to a pocket universe (second from left) that begins to grow. However, the growth rate of the pocket universe is smaller than the inflation rate of the original vacuum configuration and the pocket universe never catches up to it. Subsequent frames show additional pocket universes that are created by quantum fluctuations. Pocket universes also give rise to their own pocket universes.

In Fig. 7-6, the original stable vacuum configuration is inflating, increasing its volume by a factor of three per frame. The pocket universe that springs up in the second frame also grows fast, but at a slower rate, and never catches up to the original region. The reason why it grows at a slower rate is that the pocket universe is created by bubble nucleation, which is a quantum tunneling process, and quantum mechanical tunneling processes have a low probability. In contrast, the exponential growth of the stable vacuum is likely and depends on the value of the cosmological constant. When the first pocket universe is formed, the part of the original region that remains has grown larger than before the decay. Similarly, the two subsequent frames show additional pocket universes growing within the initial region. At all times, the vacuum configuration that is left is larger than both all the pocket universes combined and the original region that gave rise to them.

According to the eternal inflation model, the process continues forever, continuously generating pocket universes. Each valley in the landscape will endlessly produce an unlimited number of universes, creating a fractal pattern, so that the generation of a pocket universe that would grow to generate its own pocket universe is repeated over and over at smaller and smaller scales (Plate X). If this view is correct, each one of the estimated 10^{500} valleys in the landscape will incessantly create an infinite number of universes at an ever-increasing rate for an eternity.

The bubbles that are created in the stable vacuum configurations and that seed the pocket universes tunnel out of their original valley and populate neighboring valleys that match the new lower energy of each bubble. The quantum tunneling of a bubble to neighboring valleys of lower energy is random and occurs throughout the landscape at all times. In this way, each valley is visited an infinite number of times by all the possible decays that originate in valleys higher up in the landscape.

Curiously, the internal inflation scenario resembles the old steady-state model of the universe that British cosmologists Hermann Bondi, Fred Hoyle, and Thomas Gold of the University of Cambridge introduced in the 1950s. In their model, the universe obeys what they called the *perfect cosmological principle*, according to which the universe is homogeneous and remains unchanged over time to any observer anywhere in the universe. To accomplish this immutability of the universe in time and space, the Cambridge trio proposed that new galaxies emerge continuously within the expanding space. The spectacular verification of the predictions of the competing model—the big bang theory of the origin of the universe—with the detection of the microwave background radiation by Penzias and Wilson in 1963 and, later, with the COBE and WMAP observations, showed that the steady state model was incorrect.

At any time, the multiverse contains an enormous number of expanding bubbles within bubbles, with each bubble starting out with a short and rapid inflationary period, followed by a big bang. The inflationary period assures that each universe ends up almost perfectly flat, like our own universe. Each pocket universe rests on its own valley in the landscape, with its own value of the cosmological constant, and its own set of the laws of physics. Our own universe sits in one of the bubbles, which is in turn nested in a large network of larger and larger bubbles. Our visible universe, some twenty-seven billion light-years across, is a very small region within one of these nested bubbles (Fig. 7–7). We can't even see our entire universe, much less neighboring bubbles. These bubbles will remain out of our reach forever.

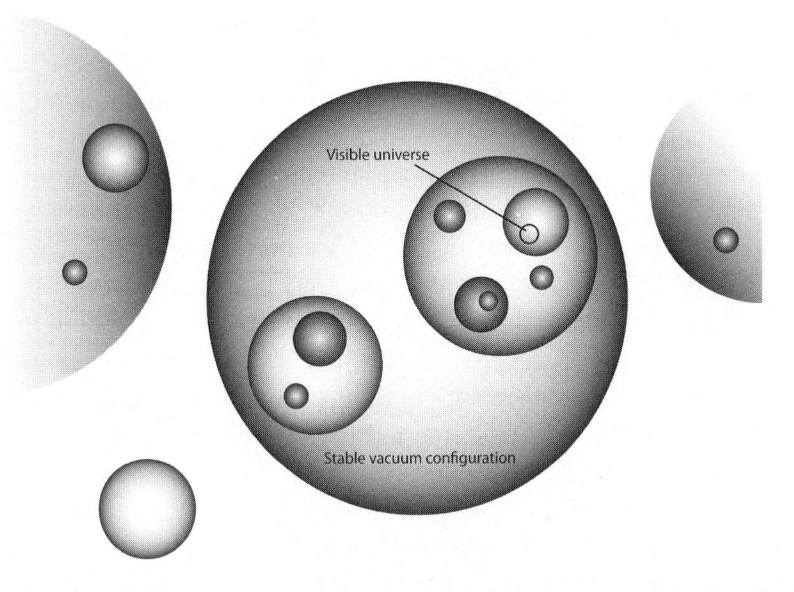

Figure 7-7. Our visible universe, some twenty-seven light-years across, is a region in one of the nested bubbles that came out of one of the 10^{500} stable vacuum configurations in the landscape. We can only see as far as the boundaries of the visible universe. The entire bubble where our universe exists is not all visible to us. Other bubbles are way out of our reach. The situation is the same for any hypothetical observers in any other universe.

If the eternal inflation model is correct, our big bang was nothing more than the most recent quantum tunneling to our valley in the landscape where a new string configuration gave rise to a new set of Calabi-Yau shapes for the hidden dimensions. This particular configuration gave us our cosmological constant and our set of the laws of physics. Far in the future, another quantum fluctuation will take place in our bubble and the universe as we know it will end, to give way for another universe with its own set of laws.

Eternal inflation tells us that an infinite number of universes will randomly and spontaneously sprout out of each one of the stable vacuum configurations, and that the rate of pocket universe production increases with time. Going back in time, has the creation of pocket universes been happening for eternity? The idea of a universe that has existed forever is a very attractive one for many people because it solves the problem of the creation of the universe. In 1994, Vilenkin together with Arvind Borde of Brookhaven National Laboratory studied this question under the framework of string theory and came up with a preliminary and tentative answer.[6] According to their approximate calculations, if the universe obeys certain specific conditions, the universe must have a beginning, although it can continue existing forever. Further examination by Borde and Vilenkin, however, made them realize that one of the assumptions they had made, valid at the classical level, did not apply to the quantum case and conclud that the question of the origin of the multiverse was still open.[7] I think that our knowledge of the existence of these specific conditions in our universe is too rudimentary at this time to be able to obtain a reliable answer.

The string theory landscape and its equivalent approach of eternal inflation tell us that if this model is correct, all the estimated 10^{500} valleys house a universe with its own cosmological constant and set of physics laws. These universes are real and underwent a similar early evolution. We know that one of them has the conditions that allow life to exist. Are most of the other universes hospitable to life or is life a rare phenomenon? We can ask the equivalent question: Is the extremely small but nonzero value of the cosmological constant common throughout the other universes or is fine-tuning still required to select our specific value?

EXPLAINING THE FINE-TUNING

Our pocket universe occupies a valley in the landscape that provides a near perfect cosmological constant for life to exist. As far as we know, it is perfect, since it did lead to life. Other values of the cosmological constant that are very close to our value could also work. But deviations larger than one order of magnitude in either direction would lead to lifeless universes. The cosmological constant must stay in that extremely narrow range around the very small value that we have measured for our universe.

The multiverse with 10^{500} pocket universes should have plenty of valleys with a value of the cosmological constant in that range. How many? About one in 10^{120} should lie in the right range. That's 10^{380} pocket universes with the right cosmological constant for life. While that is a huge number of universes, it's still a tiny fraction of all the pocket universes that are supposed to exist, since there would be nearly 10^{500} universes with the wrong cosmological constant.[8] The probability of selecting one of the 10^{380} universes is extremely small.

It's sometimes hard to visualize overwhelming odds, and these odds are extreme. State lotteries are a prime example of overwhelming odds. On occasion I illustrate to lottery players how unlikely it is to win the top prize. Typically, the odds of winning a multistate lottery are 1 in 50 million. Suppose that you take 50 million marbles and dump them all in a football field. The marbles will roughly cover the whole field one layer deep.[9] Only one of the 50 million marbles covering the field is worth the top prize and now you have one chance to select that marble and win the prize. Why bother, right? And those odds are only 1 in 50 million, or 1 in 5×10^7. If you are in the business of selecting universes from the landscape, your odds are 1 in 10^{120}.

We seem to be back where we began. Our universe appears to be designed for life and the key to life lies with the value of the cosmological constant. If the universe is unique, it is extremely hard to understand how—on its own and by chance—it ended up with the perfect conditions that made life possible. But with the development of M-theory and the discovery of its innumerable solutions, some scientists saw a way. Our universe, with its perfectly tuned cosmological constant, is only one of many other universes in a vast landscape of pocket universes, each with its own value of the cosmological constant. Using basic concepts of general relativity and quantum mechanics, Bousso and Polchinski, as well as other scientists, were able to find plenty of possible pocket universes with the right value, but selecting one of those turned out to be a very unlikely event. Do we now turn to a designer to make the unlikely selection?

There is another way to look at all of this. Life evolved on Earth but not on Jupiter or Pluto because Earth is the only planet in the solar system that can sustain it. No one designed Earth for us. There are billions of stars in our galaxy, one of billions of other galaxies in our visible universe. Many of those billions of stars have planetary systems, and astronomers have begun to discover them (Plate XI). Most of those planets are inhospitable to life because they are too close to their suns and are too hot. Others are too large and have retained dense atmospheres that absorb most of the light from their stars. Still others might be too small and do not have enough mass to retain any atmospheres at all. A small fraction will have the right conditions for life. Earth is one of those planets, and life arose here. Life exists where it can. This kind of argument is called *anthropic*.

The problem is that, while this argument is reasonable, we don't learn anything from it and, superficially, it even appears to be trivial, bordering on being a tautology. However, some physicists think that a more careful examination would show that applying this reasoning to the multiverse could have profound implications. Only a tiny fraction of all the pocket universes in the multiverse have the right cosmological constant that makes possible the evolution of life. Our universe is one of those and life arose here. Many universes with every kind of possible value of the cosmological constant exist with their own specific laws of physics. Life arose in at least one of the 10^{380} pocket universes that had the right laws of physics. No one designed our universe for life just as there is no need for a designer to design trillions of universes in which the laws of physics lead to chaos. Life was not planned for, it simply happened in our universe and it likely happened in most if not all of the 10^{380} universes that may exist in the multiverse and that have cosmological constants in the right range. And life didn't happen in the majority of the universes.

The anthropic argument does offer a solution to the delicate equilibrium that we observe in the universe. But it is a solution that many physicists find very difficult to accept. Other physicists, such as Nobel laureate Steven Weinberg of the University of Texas, think that this argument is the only answer. In 1987, before the experimental determination of the cosmological constant was made, most physicists thought that this value should be exactly zero. Weinberg predicted then that for galaxies to form in our universe, the cosmological constant could not be zero.[10] But it couldn't be large either. According to his calculations then, the cosmological constant should be such that it "would have to cancel out to 120 decimal places." That was exactly the value obtained from the observations of the variation in the rate of the expansion of the universe made by the supernova studies of 1998.

But Weinberg arrived at his conclusion with some reluctance. "If such a cosmological constant is confirmed by observation," he wrote in 1992, "it will be reasonable to infer that our own existence plays an important part in explaining why the universe is the way it is."[11] But he added: "I hope that this is not the case. As a theoretical physicist, I would like to see us able to make precise predictions, not vague statements that certain constants have to be in a range that is more or less favorable to life." Some physicists still hope that a more encompassing theory could one day determine unequivocally the laws of physics, the constants of nature, and the cosmological constant.

An all-encompassing theory is what string theorists claim M-theory is. But the predictions of M-theory are far from being proved. In fact, the precise formulation of the theory is still unknown and many of the solutions are only approximations. Moreover, the theory doesn't offer any predictions that can be tested with current technology. The theory promises much, but much needs to be done to it before it can be considered as one of the ironclad theories of physics, like quantum mechanics or general relativity. However, one thing seems irrefutable: the nonzero value of the cosmological constant.

Even if M-theory turns out to be incorrect, the idea of the multiverse is likely to stay with us. (It is part of many of the approaches that I discuss in the remaining chapters.) Here, I'd like to briefly describe one interesting approach at this idea of the multiverse that the physicist Lee Smolin of the University of Waterloo has proposed.

Cosmological Natural Selection

Taking a clue from the success of the evolutionary theory in biology, Smolin proposed in 1987 that the multiverse could also be controlled by a *cosmic natural selection*. Smolin reasoned that in biology, natural selection provides a solution to a problem of explaining improbable complexity. "To my knowledge," he has recently written, "only in biology do we successfully explain why some parameters—in this case the genes of all the species in the biosphere—come to be set to very improbable values, with the consequence that the system is vastly more complex and stable than would be for random values."[12]

How does cosmic natural selection work? In Smolin's model, universes are born when black holes collapse. Black holes are formed when massive stars spend all their nuclear fuel and begin to rapidly contract. Without fuel, a star is not able to withstand gravity. In the case of less massive stars like our sun, the contraction heats up the core of the star, igniting it again, and giving the star a second lease on life. The new radiation of energy from the core expands the star and makes it into a red giant. When this energy is exhausted, the star collapses again, reignites once more, and the star shines as a white dwarf. When that energy is used up, the star dies, ending its life as a burned-out black dwarf. The life of massive stars—those with masses larger than 3 solar masses—is somewhat simpler. The gravitational pull of such a star is so great that, once the nuclear fuel is spent, nothing stops the initial collapse and the star becomes a black hole (Plate XII).

A black hole is formed when the entire mass of the star collapses to a small size, increasing the strength of its gravitational field to a point that not even light can escape. A star with mass equivalent to 3 suns would have to be squeezed down to a sphere with a diameter smaller than 18 kilometers (11 miles) before it becomes a black hole. For a 4 solar mass star, the sphere is 24 kilometers (15 miles). This sphere is called the *event horizon*, the boundary of the black hole.

You likely recall that Einstein's theory of special relativity says that light always travels at the same speed. How can gravity, regardless of how strong, slow down light? According to general relativity, gravitational fields slow down time. The stronger the field, the slower time flows. At the event horizon, the gravitational field is so strong that, as seen from the outside, time is stretched by an infinite amount, which is the same as saying that in the vicinity of a black hole, time doesn't flow. From outside the event horizon, light takes an infinite amount of time to leave the star. Since the star is still emitting light, light must be trapped inside the black hole.

General relativity also says that the collapse of a star into a black hole tears the fabric of spacetime. All the matter of the star is pulled together into a single point of infinite density, a *singularity*. At the singularity, space ends and time ends and general relativity breaks down. General relativity is not equipped to deal with the behavior of small objects. That's the realm of quantum mechanics. And quantum mechanics doesn't deal with objects that have densities approaching infinity. One way out of this problem is to combine the equations of general relativity with quantum mechanics into a theory of quantum gravity that would tell us precisely what really happens when a black hole approaches the singularity. However, no one has fully succeeded in accomplishing the marriage between these two theories. Of course, that's exactly what M-theory is supposed to do, marry gravity or general relativity with quantum mechanics. However, in M-theory, actual calculations of difficult problems such as this have not been performed yet due to the tremendous complexity of the equations. Back in 1974, long before M-theory or superstring theory existed, John Wheeler of Princeton University performed some preliminary calculations in quantum gravity and proposed that a black hole would avoid the singularity, stop collapsing, and start expanding.[13] According to Wheeler's conjecture, a new spacetime expands out of the black hole. More recently, approximate calculations in string theory have produced similar results.[14] A promising new approach to quantum gravity called *loop quantum gravity* has been introduced since the 1980s and some calculations with this model show that cosmological singularities do bounce into expansions.[15] Work on obtaining similar results with black hole singularities is only starting.

Smolin proposes that these calculations are describing the actual behavior of black holes and builds his model on this premise. In his model, the new region of expanding spacetime in the blackhole becomes a new pocket universe. Since the black hole is inaccessible from outside of the event horizon, the new universe remains invisible from our world. In Smolin's model, not only is every single black hole in our universe generating new universes, but our own universe is the result of the collapse of a black hole in another universe.

How does Smolin's cosmological natural selection work? Smolin assumes that the laws of physics of the new universe that sprouts from a black hole are *slightly different* from the laws of physics in the parent universe. These differences must be small so that the transformations are simple enough to have been produced at random. This assumption assures that the new universes would have properties with many common traits. The number of universes that are born out of a universe depends on the laws of physics at work in the parent universe. In very young universes, not enough time would have passed for many of its large stars to go through their life cycles and end as black holes. In old universes, many of their larger stars might already be black holes that have already given birth to their descendants. As in biological systems, universes that are neither too young nor too old would be the ones with the largest progeny. There is then a selection pressure: since the universes with laws of physics that favor the formation of black holes would have the most offspring, they would eventually take over and dominate.

Smolin has used his model to make predictions that can be tested experimentally. One such prediction is that if all universes, including our own, are produced by the selection process that resulted in the maximization of black hole production, the production rate of black holes in our universe is optimal and can be, in principle, measured. The processes that affect black hole formation are nucleosynthesis, galaxy formation, and neutron star dynamics. Except for galaxy formation, all of these processes are relatively well understood.

Even when we have a more complete understanding of these processes, including galaxy formation, and scientists are able to obtain reliable data on the rate of production of black holes, how do we know that this rate is optimal? We will only have data for one universe. How do we determine what is optimal with only one sample? Smolin has proposed that if our universe is a typical member of those universes created by this selection mechanism, its laws should be maximized for black hole production.[16] Thus, any modification of its laws should result in a decrease in the production rate of black holes. These concrete predictions can be checked experimentally and would allow for the verification or rejection of the prediction without the need to know the specific laws of physics in other universes.

A difficult problem to reconcile is the question of the loss of information through the black holes that generate new universes. There seems to be general agreement in the physics community now that information is conserved in the universe. The cosmological natural selection model violates that conservation principle.

Linde and Susskind have argued that the cosmological natural selection model is incorrect because the huge number of universes created in eternal inflation would overpower the number of universes produced by black hole decay. As Smolin has pointed out, although the inflationary theory predictions have been strongly validated, the same cannot be said of the eternal inflation model. As a result, the rate of universe production through eternal inflation is not reliable yet.

And so the debate continues. As has been the case in many instances in the history of science, there are competing theories struggling for survival. There are other approaches still. An interesting possibility is that of a universe that has existed forever, oscillating between a big bang expansion and a collapse that leads to a new big bang and does away with the problem of origins. But does it do away with the fragile equilibrium, the apparent fine-tuning of the cosmological constant?

Chapter 8

Eternally Oscillating Universes

A Very Brief History of Time

Our knowledge of the laws of physics has given us an account of the origin of our universe much more detailed than what the Babylonians, Egyptians, early Greeks, or Israelites ever dreamed of possessing. Whether our universe is all that exists or is simply a pocket universe in a vast landscape of many universes, it now seems almost certain that the big bang was the event that created our laws of physics as well as the matter and force particles that make up all that we observe. And we now have a very detailed description of how it all happened.

In the beginning, there was no light. There was only vacuum energy, the potential energy of spacetime with no matter in it. For the first 10^{-43} second, there was complete simplicity and total symmetry. This unimaginable minute fraction of one second is called the *Planck era*;[1] at that time, the entire universe was confined to a region 10^{-33} cm across, a distance known as the *Planck length*. The vacuum energy, the potential energy of spacetime, existed in a symmetric inflation field configuration. This inflation field triggered a period of extremely rapid expansion. According to the inflationary theory developed by Alan Guth, Andrei Linde, Andreas Albrecht, and Paul Steinhardt—as described in chapter 4—inflation lasted for 10^{-35} second, during which the universe doubled in size every 10^{-37} second. In this short period, the diameter of the universe doubled one hundred times. When inflation ended, the universe had a perfectly flat geometry and was homogeneous with slight temperature variations produced by quantum fluctuations.

At 10^{-35} second after the big bang, the inflationary energy that expanded the universe 10^{100} times decayed into a hot plasma of radiation and elementary particles at a temperature of 10^{28} K. Gravity had emerged earlier, breaking the total symmetry that existed during the Planck era. Inflation has ended and the big bang expansion that would drive our universe today took over. Now, the color force that would later underlie the strong nuclear force appeared as an independent force.

When the universe was a 10^{-11} second old, it had cooled down to about 10^{15} K and the weak and electromagnetic forces broke away by the process of symmetry breaking. From then on, the four forces of nature known today began their existence independently of each other. Quarks and antiquarks, electrons and positrons (antielectrons), gauge bosons, and all the Standard Model matter and force particles interacted with each other at very high energies. All these interacting particles formed the intense radiation of that epoch with the matter particles participating in the radiation as they continuously popped in and out of existence. These particles were closely packed and the time between collisions was very short. Although photons move at the speed of light while all matter particles had to have subluminal speeds, the frequent interactions of the photons with matter caused them to constantly change directions. As a result, matter particles and radiation remained together in the form of a plasma in thermal equilibrium.

At 10^{-5} second, as the universe expanded and cooled to a temperature of 10^{12} K, almost all of the quarks and the antiquarks annihilated each other during their energetic collisions that had been taking place. But they didn't start out in equal numbers. There was an imbalance, with slightly more matter than antimatter particles. The small number of quarks and gluons that remained became confined into protons and neutrons, the particles that would eventually form the nuclei of atoms. But the energies were still too high for nuclei to form. The universe at this time consisted of a soup of hadrons (protons, neutrons, and all the particles that participate in the strong nuclear force), along with radiation.

When the universe was a millisecond old—at 10^{-3} second—most of the interactions involved the electromagnetic and the strong nuclear force. As it happened with the quarks and the antiquarks, the leptons and the antileptons annihilated each other, leaving a relatively small number of excess leptons. These leptons—electrons and neutrinos—along with the nucleons formed with the excess quarks were the seeds for the formation of atoms that would make up the stars, planets, and galaxies that we see in our universe today. The origin of the matter-antimatter asymmetry that made possible the formation of structure in the universe is still an open question.

By the time the universe was one second old, the temperature of the universe was low enough for protons and neutrons to come together and form the nuclei of the lightest elements: hydrogen and helium. This process of nucleosynthesis continued for ten minutes, at the end of which there were ten nuclei of hydrogen for every nucleus of helium, the ratio observed today.

During this early epoch, most of the energy of the universe was in the form of radiation. This radiation-dominated epoch lasted for about 75,000 years. The quantum fluctuations that generated temperature variations at the end of the inflationary period produced variations in the density of the nuclei formed in the nucleosynthesis process. After 75,000 years of expansion and cooling, the temperature of the universe was low enough for gravity to start pulling more matter toward these higher-density regions. The formation of these clumpy regions that were the seeds of future galaxies marked the beginning of the matter-dominated epoch of the universe.

Some 300,000 years after the big bang, atoms began to form when electrons linked up to nuclei through the electromagnetic force. When all the electrons became attached to nuclei to form neutral atoms, the photons, which had been interacting with charged matter, began to uncouple, embarking on an independent existence throughout the universe. The universe became transparent to light. These primordial photons would travel completely undisturbed along straight lines throughout space, cooling as the universe expanded. When they were first liberated from the plasma, their temperature was 3,000 K. Today, their wavelengths have shifted to the microwave region, with a temperature of 2.725 K. These photons form the cosmic microwave background radiation that the COBE and WMAP satellites detected (Plates VIII and IX).

One hundred million years after the big bang, the first stars formed and began to reionize hydrogen (Plate XIII). The first galaxies began to appear about 300 million years after the big bang. The regions of higher density that began to form when the universe was 75,000 years old grew into stars. Gravity acted to clump these stars into groups that became the earliest galaxies (Plate XIV). Galaxy formation took between 20 and 100 million years, depending on the size of the galaxy. Our Milky Way galaxy took about 100 million years to fully form.

One billion years after the big bang—near the current limit of our direct observations—large galaxies formed and reionization was complete. Star formation peaked and began to decline at 4 billion years.

Most of the early stars didn't live long, ending their lives in powerful explosions called supernovas (Plate I). These explosions spewed out carbon, oxygen, silicon, and iron. New stars were born out of the matter released by the explosion as well as the interstellar matter already existing in the galaxies. About 5 billion years ago, in one such galaxy, one with four arms in the form of a spiral, a yellow star was born among billions other stars. And in one small clump of matter in the gas that formed that star, a clump that contained light and heavy elements began to cool off to become the planet we call Earth.

BRANE WORLD

Our knowledge of the laws of physics, supported by the precise observations obtained by COBE and WMAP, has given us a glimpse at how the universe evolved from the Planck time to the present. We still don't know exactly what happened at the moment of the big bang. It still isn't clear whether the eternal inflation model advanced by Linde, Susskind, and others has been happening for an eternity or if it simply transfers the problem of the origin of our universe to the origin of the multiverse. Lee Smolin's idea of universes coming out of the decay of black holes could support an unlimited chain of universes that will continue to be born and evolve under the rule of natural selection. Going back in time, however, we end up with a primordial first universe that gave rise to it all. How did it come about? At this stage in the development of the eternal inflation and cosmological natural selection models, the problem of origins remains unresolved.

In 2002, Paul Steinhardt of Princeton University and Neil Turok of Cambridge University proposed a model of a universe with no beginning and no end.[2] In their Cyclic Model, the big bang is not the beginning of time but the transition from an earlier phase in the evolution of the universe, one of an infinite number of phases that start with a bang and are followed by a slow expansion, like the one we are experiencing now. In each cycle, the expansion slows down, stops, and turns into a slow contraction, ending with a big crunch that becomes the next big bang that fuels a new cycle.

Steinhardt and Turok based their model on the work of Edward Witten and Petr Hořava at the Institute for Advanced Studies in Princeton.[3] Recall that the fundamental ingredients of M-theory are multidimensional branes with up to eleven spacetime dimensions. The Hořava-Witten or HW theory describes an eleven-dimensional universe that consists of two parallel branes, each containing nine spatial dimensions and one of time that enclose a region of eleven spacetime dimensions. This higher-dimensional region of spacetime is called the *bulk* (Fig. 8–1). Each brane has its own set of particles and forces that are, in general, different from those of the other brane.

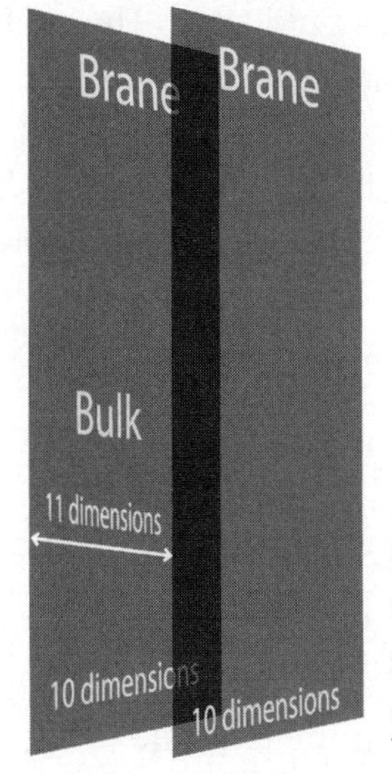

Figure 8-1. In HW theory, the eleven-dimensional universe exists in two ten-dimensional branes with nine spatial dimensions plus time that bound the eleven-dimensional bulk.

Since our world doesn't have ten spatial dimensions, the next step was to come up with a realistic model that could describe our universe. Hořava and Witten assumed that six of the ten spatial dimensions are compactified into a Calabi-Yau shape, leaving a brane with three spatial dimensions—a three-brane—enclosing a five-dimensional bulk with four spatial dimensions plus time. The four spacetime dimensions of each brane would match those of our world.

Hořava and Witten assumed that all the Standard Model particles reside on one of the three-branes (which would correspond to the one we inhabit). Quarks, electrons, neutrinos, and all matter particles as well as their force carriers, such as the photon or the gluon, are described by open strings whose ends reside on the three-brane. The two ends, confined to the brane, can move along the brane. Since the vibrational modes of an open string are particles, the two ends of the string are particles that move and interact only on the brane. However, the graviton, the carrier of the gravitational force, is a closed string with no ends that could be attached to branes. The graviton is allowed to travel through the bulk, the higher-dimensional spacetime around the branes.

What kinds of particles would the other brane contain? In the HW model, the other brane would contain all the other particles allowed by the theory that we have not detected in our universe, such as the supersymmetric partners. Gravity would be present everywhere, in each one of the branes and in the bulk.

In 1999, three years after the HW theory was published, Henry Tye and Zurab Kakushadze of Cornell University showed that the HW scenario, with Standard Model force and matter particles living inside a brane and gravitons in the higher-dimensional bulk, could be a viable description of the universe.[4] They called their scenario *brane world*. Later, Tye and two other Cornell physicists proposed a brane world cosmology; soon after several other researchers followed suit.[5]

THE UNIVERSE NEXT DOOR

That same year, Lisa Randall of Harvard University and Raman Sundrum of Johns Hopkins University proposed a brane world model based on the HW theory that attempts to explain the extreme disparity between the strength of the force of gravity and the other three forces of nature.[6] As in the HW scenario, the universe in Randall and Sundrum's model consists of two four-dimensional branes that bound a five-dimensional bulk (Fig. 8–2). One of the branes, called the *Weak brane*, contains all the Standard Model particles along with the Higgs particle responsible for breaking the electroweak symmetry. The name of this brane makes reference to the electroweak mass scale, the mass scale of the Standard Model particles. The other brane, called the *Planck brane*, contains particles different from the ones observed in our universe. The name refers to the *Planck scale*, the mass scale associated with gravitational interactions, which, as we'll see shortly, are more important on this brane.[7] The region in between the two branes, the five-dimensional bulk, contains the graviton that arises from a closed string and must be confined to the bulk.

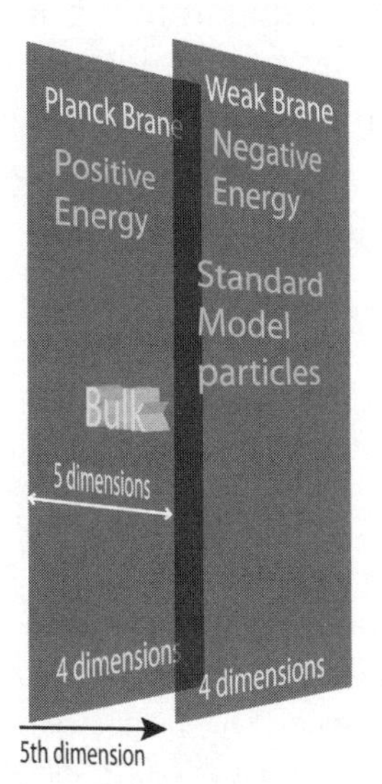

Figure 8-2. In Randall and Sundrum's model, the universe consists of two parallel four-dimensional branes that enclose a five-dimensional bulk. One of the branes, where our visible universe resides, contains all the Standard Model particles as well as the Higgs particle.

The two branes, the Planck brane and the Weak brane, carry energy, but their values are very different on each brane: the Planck brane has positive energy, whereas the Weak brane has negative energy. This difference affects the behavior of the graviton throughout the bulk—which also carries energy—and in the vicinity of the two branes in a very specific way. According to Einstein's general relativity, energy induces gravity and gravity curves space and time. Randall and Sundrum solved Einstein's gravity equations for this bulk-two brane geometry and found that the spacetime through the five-dimensional bulk has an extremely large curvature near one of the branes and almost no curvature in the region close to the other brane.

One way to illustrate this variable curvature of the bulk is by plotting the gravitational field strength through the bulk, across the fifth dimension (Fig. 8–3). General relativity says that the curvature of spacetime depends directly on the strength of the gravitational field. Therefore, near the Planck brane, where the spacetime curvature is maximum, the field strength is largest. As we move away from the Planck brane, along the fifth spacetime dimension, the strength of the gravitational field drops rapidly, becoming weak as we approach the Weak brane, where the curvature is minimum.

Gravity interacts with the elementary particles residing on each one of the branes in different ways. However, since interactions with elementary particles are governed by the laws of quantum mechanics, we need to approach gravity from a quantum mechanical point of view. And when we do that, when we combine gravity with quantum mechanics, we get the graviton, the carrier of the gravitational force, a particle with a diameter of the Planck length, 10^{-33} centimeter.

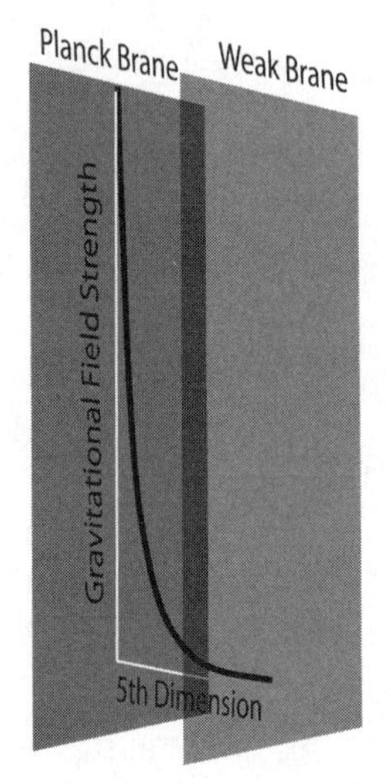

Figure 8-3. The strength of the gravitational field is strongest near the Planck brane, where the curvature is extreme. The field strength drops fast as it moves across the fifth dimension, away from Planck brane and toward Weak brane. The extraordinarily sharp exponential drop is a consequence of the energy of the branes and the bulk.

Things aren't that simple, however. Combining the Standard Model with gravity results in a theory with a ridiculously large disparity in the energy scales. At one extreme is the energy of the Standard Model particles, which is characterized by the electroweak scale energy, the energy at which the electroweak symmetry is broken. At the other extreme is the Planck scale energy that determines the strength of the gravitational interactions with other elementary particles. On the Weak brane, where the Standard Model particles reside, gravity is very weak. From Newton's law of gravitation we know that the strength of the force of gravity is inversely proportional to the square of the energy. Since the strength of gravity in our world is so small, the energy is very large, sixteen orders of magnitude larger than the electroweak energy. As noted in chapter 6, physicists refer to this large disparity in the energies as the *hierarchy problem*. The disparity in energies results from the weakness of the gravitational force when compared to the other forces of nature. The hierarchy problem is also expressed as the large disparity between the force of gravity and the other three forces.

Recall that the hierarchy problem is unresolved in the Standard Model. One of the appealing features of Randall and Sundrum's model is that it provides a solution to the hierarchy problem. The solution came about because of the specific warping of spacetime throughout the bulk with a very large curvature near the Planck brane that results in a very strong gravitational field. You'd expect to find a large number of gravitons near the Planck brane, where the gravitational field is strong, and relatively very few near the Weak brane, where the field is weak. Because the Heisenberg uncertainty principle prevents us from specifying precisely the location of the graviton (or of any elementary particle), quantum mechanics prescribes a way of determining the probability of finding the graviton at a given location during a given time interval. The quantum mechanical prescription is called the *probability function* of the graviton. The probability function of the graviton has a relatively large value near the Planck brane, where gravity is strong, and decreases rapidly as you move through the bulk, away from the Planck brane and toward the Weak brane, where the Standard Model particles reside. The shape of the probability function is exactly the same as the shape of the field strength in Fig. 8–3. The specific form of the probability function is a consequence of the energy of the branes and in the bulk. The probability function, maximum near the Planck brane, drops exponentially to a very small value very quickly; this value approaches the value that it has near the Weak brane.

The dramatic drop in the strength of gravity as we move away from the Planck brane determines the interaction of the graviton with any other particles. Near the Planck brane, where the graviton's probability function is large, the energies of the graviton's interactions with the particles that inhabit that brane are large. However, near the Weak brane, the graviton probability function is extremely small; the energies of the graviton's interactions with the Standard Model particles that exist on the Weak brane are very weak—sixteen orders of magnitude weak—and that is what we observe in our universe.

The hierarchy problem is solved because the extreme curvature of the gravitational field strength throughout the bulk, as you move from the Planck brane to the Weak brane, provides the scaling factor. Gravity strength is on a par with the other forces of nature on the Planck brane but is very weak everywhere else, in particular, on the Weak brane where our universe resides.

Randall and Sundrum's calculations show that the probability function of the graviton—or equivalently, the strength of the gravitational field—is such that it does not require a large separation between the two branes. In fact, it requires that they be close. If the branes were too far apart, the force of gravity, already diluted at the location of the Weak brane, would be diluted too much, violating Newton's inverse square law. How far is the other brane? Nima Arkani-Hamed of Harvard University, Savas Dimopulos of Stanford University, and Gia Dvali of New York University suggested that the two branes are separated by a gap of a millimeter. In the Hořava-Witten model, the two brane worlds are only 10^{-28} centimeter apart; the other universe is right on top of us.

THE CYCLIC MODEL

As Randall and Sundrum showed, the separation of the two branes is determined by matching the strength of the gravitational force with the strength of all the other forces existing on each brane. In their model, the two branes are static and the strength of the forces remains constant. Their model is not a cosmology, but rather a mechanism to solve the hierarchy problem.

Steinhardt and Turok's cyclic model is a cosmology. In their model, the two branes are not static but repeatedly collide and recoil in cycles of several trillion years (Fig. 8–4). Each collision of the two branes is a new big bang that fills the universe with matter and energy, and starts a new cosmological cycle. After each collision, the branes separate and stretch exponentially so that the matter, radiation, and all the galaxies in the universe become diluted at an ever-increasing rate for a trillion years. The separation of the branes slows down, stops, and reverses direction. During the approach, the branes stop stretching but never contract so that matter and energy in the universe remain diluted. Quantum fluctuations create wrinkles in the branes so that when they finally collide, the collision doesn't happen everywhere at once. The regions that protrude more on each brane collide first. Since each collision of the branes creates matter and radiation, different regions begin to expand at slightly different times. These minute differences become the seeds for the formation of structure in the new universe.

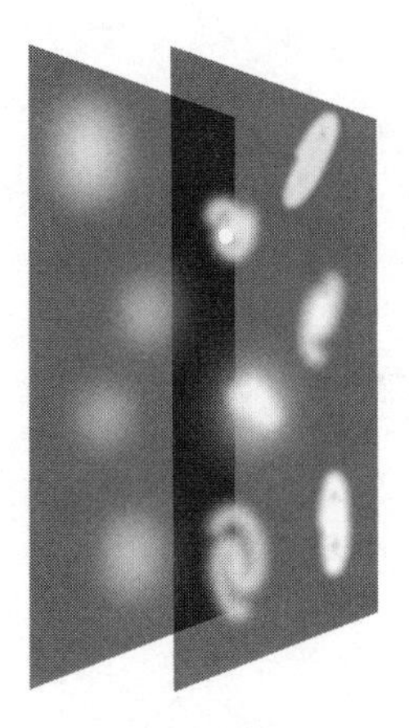

Figure 8-4. In the cyclic model, the universe consists of two oscillating branes that collide at regular intervals of several trillion years. Each collision is a new big bang that starts a new cosmological cycle. Our observable universe resides in one of the branes.

As with the HW model, the three-branes in the cyclic model reside in the eleven-dimensional spacetime of M-theory, with six of those dimensions curled up into a Calabi-Yau shape with the right properties for the existence of the observed elementary particles and forces. The branes exist in a four-dimensional spacetime—three large spatial dimensions plus one of time—and oscillate along the fifth spacetime dimension.

The two branes collide and separate like two bouncing balls hanging from threads attached at the same point. The springlike force that acts on the branes as they oscillate is controlled in part by dark energy, the energy of the vacuum, or Einstein's cosmological constant. As I explained in chapter 6, dark energy is the same as repulsive gravity. It accelerates the expansion of the universe when the galaxies separate to a large enough distance so that regular attractive gravity (which depends on distance) decreases to a very small value and repulsive gravity (which does not depend on distance) becomes dominant and takes over.

The model is built on the assumption that dark energy decays at some point during every cycle. Once we accept this assumption, the rest of it follows from well-known principles of physics. The assumption of the decay of dark energy allowed Steinhardt and Turok to derive the rules that control the behavior of their model of the universe. The springlike force that drives the oscillation of the branes depends on a specific shape of the energy density of the universe and is driven by both dark energy—repulsive gravity—and by regular attractive gravity. Steinhardt and Turok propose that the energy density has the form shown in Fig. 8-5.

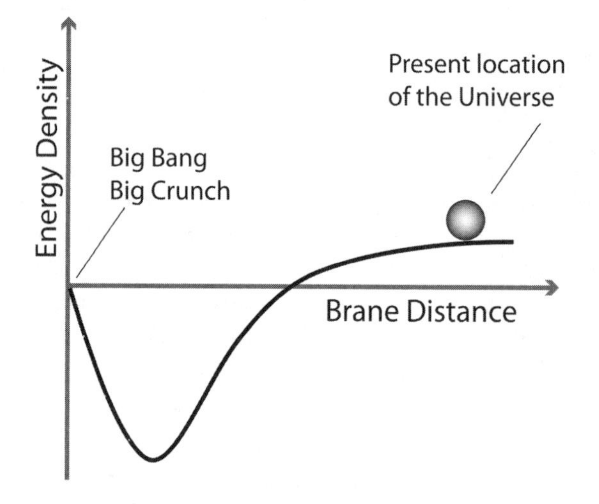

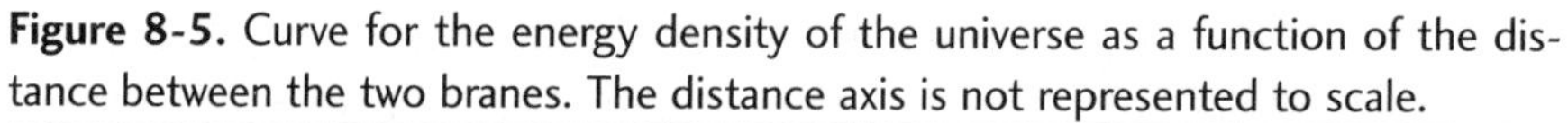
Figure 8-5. Curve for the energy density of the universe as a function of the distance between the two branes. The distance axis is not represented to scale.

To follow the evolution of the universe for one complete cycle, we can start at stage 1, just after the big bang (Fig. 8-6). The oscillation of the branes is the same as the motion of the ball along the energy curve. The branes—almost perfectly flat now—begin to rebound and contain no matter or radiation coming from the previous cycle. New matter and radiation are infused into the two branes by the tremendous energy of the collision. Quantum fluctuations occurring during the contraction phase, just before the big crunch, generate density variations that are amplified into the stars and galaxies of the new phase of the universe. The branes start to move away from each other at a fast rate and the energy density dips sharply, reaching a minimum value very quickly. As the branes move apart, the universe within one of the rebounding branes expands and cools, forming the first elements that eventually clump into stars, which later group themselves into galaxies, following the same evolution described by the inflationary and big bang theories (stage 2 in Fig. 8–6).

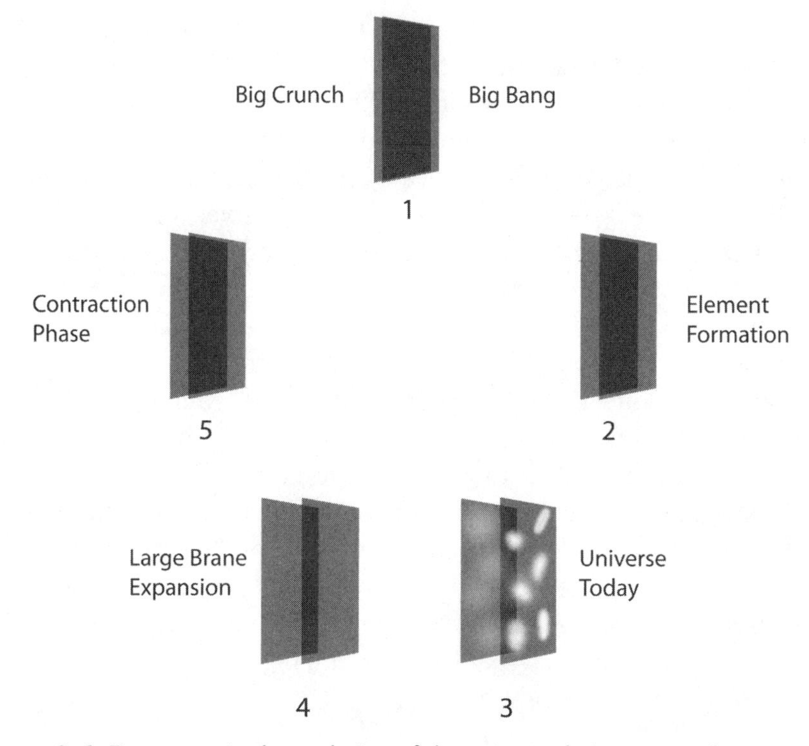

Figure 8-6. Five stages in the evolution of the universe during one cycle.

The two branes reach maximum separation in 10^{-25} second and after that continue moving so slowly that their separation—10^{-30} meter—remains almost constant for the next trillion years. Nine billion years after the expansion, the energy density crosses over from the negative to the positive side. The expansion of the universe has reached the point where the pull of gravity is balanced by dark energy—the repulsive gravity of energy of the vacuum. The galaxies continue to move apart due to inertia, reaching the region where dark energy dominates and accelerates the expansion. About 14 billion years after the big bang, the two branes reach our present location (stage 3).

At the present location of the universe, the slope is gentle and the separation of the branes decreases slowly (the ball rolls back slowly), while the universe expands. At this location, the positive energy density is the present value of the dark energy that is accelerating the expansion of the universe. The universe continues to expand for the next trillion years so that, as the galaxies move away from each other, the density of matter and radiation becomes so diluted that the two branes become nearly flat. During this phase each brane doubles in size a hundred times (stage 4).

As the branes slowly move closer, the dark energy that is driving the expansion reaches the axis again and decays to zero. At this point the accelerated expansion of the branes stops. When the universe crosses into the negative side of the energy density curve, attractive gravity dominates, and the expansion of the branes begins to slow down but never stops. As a result, the density of each brane remains diluted through the contraction phase (stage 5). The branes continue to move closer and the energy density curve dips. Toward the end of the contraction phase, some gravitational energy is converted into brane kinetic energy (energy of motion), boosting the energy of the branes just before the big crunch (upward segment of the energy curve).

During the contraction phase, quantum fluctuations are amplified into a scale-invariant pattern, just as it happens in the inflationary theory. These amplified quantum fluctuations become the seeds of galaxy formation in the next cycle. The two branes finally collide with each other in a big crunch that becomes the big bang for the next cycle of the eternally oscillating universe.

ETERNAL OSCILLATION

Steinhardt and Turok's cyclic cosmological model—with two three-branes separated by a tiny fraction of a millimeter in a hidden fifth dimension, eternally oscillating and colliding into a big crunch before rebounding in a big bang that generates a new universe at every cycle—is very appealing because it appears to do away with the question of origins. But does it? Although the idea of the universe in two branes is new, the concept of an eternally oscillating universe isn't. The idea dates back to ancient times. In modern times, one of the earliest models was proposed by Richard Tolman of Caltech in the 1930s.[8] Tolman proposed that the current expansion of the universe would eventually slow down, stop, and contract. The contraction doesn't end in a big crunch; instead, it bounces back into another cycle of expansion and contraction. Tolman based his model on one of Alexander Friedmann's models of the universe. As described in chapter 3, Friedmann had used Einstein's field equations of general relativity to develop models for an open universe that would expand forever and for a closed universe that would expand, slow down, and contract. Friedmann called his closed model a "periodic world."

Tolman soon realized that a periodic world would not work since it would violate the second law of thermodynamics. The second law says that for a closed system such as the entire universe, entropy rises on average. Entropy might decrease in one part of the system but, when that happens, a larger increase would take place someplace else in the system. During every cycle of an oscillatory universe, entropy would increase. The amount of radiation—which constitutes most of the energy of our universe—would increase in every cycle when stars are formed, and this radiation would be fed into the new cycle. Each cycle would then end up with more radiation than the previous cycles. In general relativity, an increase in radiation results in a higher expansion rate, which increases the time of the expansion to stop and reverse direction. If each cycle takes longer than the previous one, a cyclic universe would also require a beginning.

How does Steinhardt and Turok's cyclic model fare in light of Tolman's entropy argument? According to general relativity, it isn't the net increase in entropy that would produce a greater bounce and a longer-lasting new cycle after the big crunch, but an increase in the entropy *density* of the universe. The total entropy of the entire brane does increase, but since the branes always expand, even during the contraction phase, the expansion creates more space and the new entropy gets diluted. During the big crunch and subsequent big bang, the entropy density is still very low and the rate of expansion of the new cycle is controlled by the new matter and radiation, which, in this model, is always the same. As a consequence, the cycles always have the same duration.

A second important objection to the cyclic model is that oscillatory models of the universe require the universe to be closed, with enough mass density so that gravity can slow down and reverse the expansion and turn it into a contraction. Closed universes have positive curvature, but WMAP has shown that space has zero curvature; it is actually flat.

In the cyclic model, there is no need for the universe to have a closed geometry, because the oscillations aren't caused by the gravitational attraction. The oscillations come from the periodic collisions of the branes under the dual action of dark energy and gravity. The branes themselves are flat and expand forever. There is no need to stop the expansion and turn it into a contraction.

In Steinhardt and Turok's cyclic model, the universe consists of two branes with different physical properties, forever colliding and rebounding, oscillating in the hidden fifth dimension of the bulk. Our visible universe exists in one of the two branes. The other brane contains another universe with its own set of the laws of physics, its own particles, and its own galaxies, stars, and planets—a universe residing a small fraction of a millimeter away from us along the fifth spacetime dimension, but undetectable to our instruments, at least for now. Since gravity acts through that fifth dimension, we might be able to make use of it one day and detect the fifth dimension and the sister universe existing next to us. We won't be able to ever travel to it or send messages, but we would know that it is there.

THE CYCLIC MODEL VERSUS ETERNAL INFLATION

The cyclic model is a cosmological model. So is the inflationary model. The new inflationary model made very specific predictions about the nature of the temperature fluctuations that gave rise to the structure observed in the universe today, as well as the stars and galaxies that populate it. Those predictions were triumphantly corroborated by WMAP in 2006. The new inflationary theory is very strong because of that. Why should we pay attention to a new model, if nothing is wrong with inflation?

Most scientists would agree that the inflationary theory is a very strong theory and that it represents the correct approach to the origin and evolution of the early universe. Recall that according to the theory, a tiny fraction of a second after its origin, the universe expanded rapidly for 10^{-35} second, during which it expanded an unimaginable 10^{100} times. Very shortly after its birth, the universe settled in an unstable symmetric state called the false vacuum. The energy of this vacuum state—dark energy or the value of the cosmological constant—accelerated the expansion of the universe during the inflationary era. When the energy and mass of the universe changed back to the state with attractive gravity, the universe continued its expansion as described by Hubble law and the standard big bang theory.

However, the inflationary theory doesn't solve the problem of origins. Moreover, we don't know the properties of the universe during the first instant after its birth, before inflation began, and are forced to assume that these properties were the right ones to give rise to inflation. This fine-tuning of the laws of physics is something that scientists find very difficult to accept.

The more encompassing eternal inflation model attempts to explain the fine-tuning by proposing the existence of a vast landscape containing an extraordinarily large number of pocket universes, each with a cosmological constant corresponding to a different Calabi-Yau shape. From M-theory, we know that the number of such pocket universes is estimated to be of the order of 10^{500}. However, only a tiny fraction of those pocket universes ends up with a value of the cosmological constant that could give rise to a universe with properties that would allow the existence of life. If the cosmological constant is positive and large, repulsive gravity will expand the universe and dilute any matter in it too quickly, hindering the formation of stars and galaxies. If the cosmological constant is large and negative, there would be only attractive gravity that would pull the matter together into a collapse before stars could form. The right value for life must not be so large that stars cannot form but not too small that the universe collapses.

To solve the fine-tuning problem in the eternal inflation scenario, one must invoke the anthropic principle: our universe is one of those very few with the right cosmological constant and life arose in it. With the anthropic principle, the problem is solved, but at a high price: when the going gets tough, we give up. When we can't find the answer, we invoke the anthropic principle.

Even with the anthropic principle, the problem of origins appears to remain unresolved in eternal inflation. As I stated in chapter 7, our knowledge of the conditions of the very early universe is still too rudimentary to be able to obtain a clear answer to this question in the framework of the eternal inflation scenario.

Steinhardt and Turok see another flaw in the eternal inflation model, namely, the runaway expansion of the universe that started 5 billion years ago, when dark energy took over. According to eternal inflation, the universe will continue expanding, doubling its size every 10 billion years. After a trillion years, it would have doubled in size a hundred times and matter would be spread out so thin that no structures could form. The universe would end up as a lifeless vacuum that would exist forever. The same fate would await the minority of pocket universes with the right cosmological constant.

The cyclic universe doesn't lack its share of problems. The model doesn't explain the origin of the springlike force that causes the oscillation of the branes. That force is simply assumed to exist. The authors hope that it will be included in the formulation of M-theory, but no one has done that yet.

At this point, the model is somewhat sketchy. As Steinhardt and Turok point out, the specific mathematical details of brane collision haven't been worked out yet.[9] It is assumed that the collision is quasi-elastic, with some loss of energy that is replenished by gravity, and that it avoids the singularity that plagued all other cyclic models. But that has not been proven. The other unresolved problem with brane collision is that, although the collision doesn't take place everywhere at once, where and when it takes place, the fifth dimension of the bulk vanishes to zero, creating a singularity, and that problem hasn't been solved either.

Finally, the model doesn't offer a clear mechanism for generating the scale-invariant spectrum of radiation with the appropriate tilt. Recall from chapter 4 that a scale-invariant spectrum appears to have a random pattern but can actually be decomposed into a sum of sinusoidal waves. The inflationary model predicted such a spectrum but with a tilt, such that the waves with shorter wavelengths were produced toward the end of the inflationary period, when the decreasing energy generated smaller quantum fluctuations. WMAP-detailed observations showed that the spectrum of the early universe is indeed a scale-invariant spectrum with the predicted tilt. The cyclic model, coming after the WMAP observations, does not have the mechanism for producing this spectrum.

On the other hand, the cyclic model is the only one that attempts to explain the low value of the cosmological constant without resorting to the anthropic principle. Since the universe is assumed to have gone through many cycles already, there has been time for the cosmological constant to adjust to its present value. Some time after Steinhardt and Turok introduced their model, they came across an idea that had been advanced in 1985 by Laurence Abbott of Columbia University: the cosmological constant starts out with a large positive value and steadily decreases by tunneling through a series of small quantum barriers to lower and lower vacuum energies.[10] The time for each step increases exponentially as the cosmological constant decreases, so that the universe spends more time at the lowest values of the cosmological constant. Since the increase is exponential, the last positive step takes a longer time than all the other steps combined. As Abbott showed, the process ends there, because forays into negative territory are soon crushed by gravity.

Abbott proposed his idea for the big bang model of the universe, realizing soon that by the time the cosmological constant decreased to its present value, all matter in the universe would have diluted away. That isn't a problem for the cyclic model since it has time on its side. Steinhardt and Turok proposed that Abbott's quantum field resides not on our brane but on the parallel brane (the Planck brane in Randall and Sundrum's model). By placing the quantum field on one of the branes and not in the bulk, the tunneling steps of the field are independent of the brane cycles. Early cycles with large values of the cosmological constant are devoid of stars and galaxies, since dark energy and not matter would dominate, preventing the formation of structure. It is only during the cycles that take place when Abbott's field has tunneled to small values of the cosmological constant that matter dominates for a long enough time to allow for star formation. Steinhardt and Turok's calculations show that the universe can undergo $10^{10^{100}}$ cycles with the last value of the cosmological constant!

Abbott's field has been successfully added to the cyclic model, making it the only model able to propose a solution to one of the most important problems in cosmology at the present time, the explanation of the current value of the cosmological constant. It's a value so small that it requires cancellation of all the contributions to the vacuum energy to 120 decimal places. With so much time available since the early cycles, Abbott's field can reach this tiny value.

The proposed mechanism for the small value of the cosmological constant makes the cyclic model very appealing. However, like eternal inflation, the cyclic model does not yet offer a clear solution to the problem of the origin of the universe. Quantum mechanical considerations may preclude an infinite number of past cycles. As Brian Greene of Columbia University says, quantum mechanical fluctuations, always present, ensure that there is a probability, however small, of an interruption in the cycles.[11] Even if the probability is very small, it will eventually disturb the process by changing one of the required parameters of the brane collision, such as the need for the branes to be parallel to each other. With an infinite series of cycles, all probabilities, however small, will happen and the cycling process would end. Therefore, the cycles must have had a beginning.

Work is ongoing to solve these problems. In 2007, Paolo Creminelli of the Abdus Salam International Center for Theoretical Physics in Trieste, Italy, and Leonardo Senatore of Harvard University developed a version of the cyclic model with a smooth transition from the contracting to the expanding phase that avoids any singularities.[12] This smooth transition of the oscillating branes replaces the collision with a repulsion that forces the approaching branes to draw apart just before they collide. With their scenario, Creminelli and Senatore are able to obtain an approximate scale-invariant spectrum of density fluctuations. To be able to control the smooth bouncing of the branes, they use a new field known as a *ghost condensate* that was proposed by Nima Arkani-Hamed, Hsin-Chia Cheng, Markus Luty, and Shinji Mukohyama of Harvard University.[13] This field is a new kind of physical fluid that can fill the universe with massless particles and that, when it couples to the Standard Model particles, gives rise to the possibility of antigravity. The ghost condensate field also makes the Newtonian gravitational field oscillate, thus smoothing out the bouncing of the oscillating branes.

Creminelli and Senatore's scenario is still under development. In their first attempt, they were able to obtain a scale-invariant spectrum during the contraction phase that passes undisturbed through the bounce to the expanding phase. However, this spectrum has the wrong tilt, one that isn't supported by the WMAP observations. Their choice of the particular form of the ghost condensate is arbitrary; it was chosen specifically to avoid the singularity at the brane collision. The field is much more complicated than any required by the inflationary theory and makes it less appealing. More recently, Evgeny Buchbinder and Justin Khoury of the Perimeter Institute for Theoretical Physics in Ontario and Burt Ovrut of the University of Pennsylvania were able to obtain a scale-invariant spectrum with the right tilt using a similar approach.[14] Both approaches offer a scenario of cosmological evolution that remains under control in the transition from cycle to cycle with quantum fluctuations that pass unscathed through each bounce, offering a solution to the argument that unpredictable quantum fluctuations may stop the cycles. And both scenarios are just that, scenarios that may one day underline a more complete theory.

Scientific theories and models are not decided on appeal but on experimental proof. The predictions of the inflationary theory have withstood several very hard tests. While eternal inflation rests on the shoulders of the inflationary theory, the cyclic model was constructed on the COBE and WMAP experimental data. As such, this model doesn't contradict those experiments (the Buchbinder-Khoury-Ovrut model, for example, chooses among the possible potentials that their model generates, one with a tilt that matches the observations). But the more encompassing eternal inflation model and the newer cyclic model have not passed any new tests to date. The situation might change in a few years. Inflation and the cyclic model make different predictions about the generation of gravitational waves in the early universe. The Laser Interferometer Space Antenna (LISA), proposed jointly by NASA and the European Space Agency, is a space-based gravitational wave observatory scheduled to be launched toward the end of the second decade of this century. The mission will involve three spacecraft orbiting about 5 million kilometers (3 million miles) apart in a triangular configuration that forms an interferometer able to measure the minute separation of spheres in each spacecraft caused by the distortion of space as traveling gravitational waves pass through (Plate XV). Among the scientific objectives of the mission is the detection of gravitational waves that could have been produced during the inflationary period and that should have traveled undisturbed through spacetime—a cosmic background of gravitational waves. LISA should be able to detect gravitational waves originating from the first minute fraction of a second after the big bang, when the universe was less than 10^{-35} second old.

If LISA is launched and does detect the primordial gravitational waves, that discovery would add to the successes of the inflationary theory. Although it won't say anything about eternal inflation, it would contradict the predictions of the cyclic theory, since in this model, the density fluctuations date from the contraction phase, before the slow collisions of the branes in the big crunch that preceded the big bang. The model predicts no observable gravitational waves.

Regardless of which model turns out to be viable, both theories are still sketchy about how the initial conditions appeared. The inflationary theory requires the existence of the inflation field, but no one knows exactly how that field appears. The cyclic model requires the ghost condensate field and no one knows how it appears either. Moreover, it isn't known at this time whether the ghost condensate can exist in the framework of string theory, which is itself the underlying theory for the model.

We're not done with our search for a model that can explain the universe. We have yet to explore two interesting approaches by Stephen Hawking and collaborators that attempt to do that.

Chapter 9

A Universe without Origin

The Problem of Origins

The enormous successes of the inflationary and the big bang theories of the evolution of the early universe have brought us tantalizingly close to explaining the universe. Eternal inflation—the audacious theory that proposes that our universe is one among innumerable other pocket universes that populate a vast landscape—explains the persistent problem of the fine-tuning of the cosmological constant, but it does so by reaching for the anthropic principle.

The cyclic model of the universe agrees with many of the observations that brought success to the inflationary theory, although in this case, the model was based on the observations. The ensemble of past, present, and future universes that the cyclic model generates with the collision of two oscillating three-branes solves the fine-tuning of the cosmological constant without resorting to the anthropic principle. Recent work on the model is providing solutions to some of the main arguments against the validity of the model: the problem of the existence of the bounce from contraction to expansion and the transmission of the quantum fluctuations from cycle to cycle. The cyclic model makes a definite testable prediction that differs from the inflationary model: the inflationary theory predicts the existence of a primordial gravitational wave spectrum, whereas the cyclic model doesn't. If the cyclic approach passes this stringent test and the new ideas about a soft, smooth bounce as well as an unscathed transition of the quantum fluctuations are corroborated, the model offers a real possibility of an eternal universe.

However, neither inflation nor the cyclic models are yet part of a consistent theory of quantum gravity. When and if that challenging step happens, we may be able to find out whether the new fields required by each model—the inflation field or the ghost condensate—can be described within the framework of a quantum theory of gravity. Until then, the problem of origins will not be fully resolved.

THE ROAD TO QUANTUM GRAVITY

In principle, the laws of physics that govern the universe determine how the initial conditions of the universe evolve with time. Through quantum gravity, physicists attempt to understand this problem by applying quantum mechanics to the early universe. To do that, they need to unify quantum theory and general relativity, and that task has proved to be extremely difficult. The main roadblock is that general relativity by itself is inconsistent with the basic principles of quantum theory and cannot describe phenomena at very short length scales or over very short intervals of time. The idea behind a theory of quantum gravity is to extend general relativity to make it consistent with quantum mechanics.

In quantum theory, subatomic objects have particlelike properties as well as wavelike properties and these properties manifest themselves to observers in ways that depend on the nature of the experiment. If the experiment is set up to measure particle properties, we detect particles, and if the experiment attempts to measure wave properties, we detect the wave nature of these objects. At the root of quantum theory is the Heisenberg uncertainty principle, which says that electrons and other quantum objects cannot be described using common language or concepts such as position or velocity. Because of the uncertainty principle, "empty" space is actually filled with virtual particles that continuously pop in and out of existence.

On the other hand, in general relativity, our modern theory of gravity, objects are described by their positions and velocities and both quantities can be determined precisely. If we know those quantities along with their masses, we can use the equations of general relativity to obtain the curvature of spacetime, which will allow us to deduce the future location and velocities of those particles. In principle, we can also obtain the location and velocity of those objects at any time in the past.

General relativity is a classical theory and, at the atomic scale, particles do not behave classically. Behavior at the atomic scale is governed by quantum mechanics alone and the gravitational interactions are so small that they can be neglected altogether. However, at the much smaller scale defined by the Planck length (10^{-33} centimeter), gravity cannot be neglected. If we attempt to observe a particle with an accuracy of one Planck length, quantum fluctuations will generate an energy equivalent to the Planck mass, 22 micrograms, an enormous mass for a subatomic particle and one that would make gravity the dominant force at that scale. General relativity is not equipped to handle interactions at that scale and ends up giving nonsensical results.

General relativity says that gravity is geometry, a property of space and time. The gravitational field generated by masses curves spacetime, producing wrinkles. When we get close to the Planck length, these wrinkles are more and more pronounced and spacetime becomes bumpier. Quantum mechanical uncertainties make these wrinkles fuzzier. As John Wheeler of the University of Texas said, at the Planck scale, spacetime resembles a foam—spacetime foam, as he called it.[1] At the Planck scale, quantum physics and general relativity as we understand them today are inadequate to describe the universe, and only a marriage between the two to form a theory of quantum gravity will allow us to probe this scale. The new theory will also give us the tools to study the physics at the Planck scale during the early universe.

Loop Quantum Gravity

An interesting approach to quantum gravity is a model known as *loop quantum gravity*. This model started with the work of Amitabha Sen, a young postdoctoral fellow at the University of Maryland who, in 1982, rewrote general relativity using much simpler equations than what Einstein had originally used—a feat that made the connection to quantum theory easier.[2] With this new set of equations, Abhay Ashtekar, then at Syracuse University, came up with a complete formulation of general relativity and of the concept of spacetime in the language of quantum physics and quantum chromodynamics.[3] Shortly after, Ted Jacobson of the University of California at Santa Barbara and Lee Smolin, then at Yale University, were able to obtain equations for a quantum theory of gravity with the new formulation of general relativity developed by Sen and Ashtekar.[4] Those equations turned out to be simpler versions of the basic equations of quantum gravity that had been introduced in the 1950s by John Wheeler of Princeton University and Bryce DeWitt of the University of North Carolina. Although the Wheeler-DeWitt equation had never been solved, Jacobson and Smolin were able to find exact solutions to their form of the equations.

To solve the equations, Jacobson and Smolin needed to construct expressions that described the quantum states of the geometry of spacetime. To do that, they made use of a method developed a few years earlier by Kenneth Wilson of Cornell University to describe interactions in quantum chromodynamics using the old idea of field lines. In his method, quarks occupy the nodes of a latticelike structure while the quantized lines of force representing the quark interactions exist along the edges of this lattice. Since, in the absence of matter, field lines can close in on themselves to form loops, the method became known as *Wilson's loops*. Jacobson and Smolin were able to build the quantum states of the geometry of spacetime using Wilson's loops. The quantized loops satisfied their equations and showed that space, at the Planck scale, is discrete and can be described mathematically by a network of edges in a lattice structure.

Loop quantum gravity, as the model is called today, implies that any volume comes in discrete multiples of the Planck volume, which is the cube of the Planck length, and is equal to 10^{-99} cubic centimeter. In this model, space has a discrete, quantized structure.

In 1995, Smolin and Carlo Rovelli, then at the University of Pittsburgh, showed that the quantum loop states could be arranged in the geometrical pictures called *spin networks* that Roger Penrose had discovered many years earlier.[5] Penrose developed his spin networks to represent his idea that space was relational. The networks are geometrical figures with many sides, whose edges represent the values of the angular momenta of the different interacting particles and whose nodes correspond to the quantized units of volume. These spin networks offer a powerful computational technique.

Loop quantum gravity implies that space and time are not continuous and that the discontinuity becomes manifest at the Planck length. As Smolin puts it, "Space [is] woven from a network of loops, . . . just as a piece of cloth is woven from a network of threads."[6] Space is quantized and is composed of fundamental units of space. The quantum loop defines space; nothing exists inside or outside the loop. The model has been developed with the mathematics of spin networks and tells us that the fundamental unit of space is represented by the Planck volume.

The techniques of loop quantum gravity may have applications to the study of the early universe. Martin Bojowald of Pennsylvania State University has studied quantum states that are exact solutions to loop quantum gravity and that have the same symmetries as our universe. He has also been able to show that loop quantum gravity seems to imply that there was no initial singularity at the moment of the big bang and that time continues past the point where general relativity says it ends.[7] The model shows that the universe continues back beyond the moment of the big bang to an earlier phase. Thus, according to Bojowald's calculations, the universe arose from that earlier phase, either through a collapse of a black hole, as in the cosmic natural selection model, or through the collapse of a previous cycle, as proposed by the cyclic model.

Loop quantum gravity is still a work in progress and currently far from a complete theory. The model has some appeal because it attempts to combine quantum theory with general relativity without modifying either one. It has also been possible to draw predictions from the model about quantum gravity effects that could be observed in future measurements of the cosmic microwave background radiation.[8]

The Holographic Principle

M-theory is perhaps the best attempt at the unification of all the known forces of nature, including gravity, and, in principle, should include a consistent theory of quantum gravity. However, M-theory is still in development and its equations are still approximations. The exact equations aren't yet known. In spite of that, theorists have been trying to use the theory to gain an understanding of the structure of spacetime at the Planck level.

One promising approach now being pursued with a great deal of interest is based on an idea proposed in 1997 by a young Princeton graduate student by the name of Juan Maldacena, who is now at the Institute for Advanced Study in Princeton. Recall that in M-theory, the Hořava-Witten model describes an eleven-dimensional universe that consists of two parallel ten-dimensional branes (nine spatial dimensions plus time) that enclose an eleven-dimensional bulk. The Standard Model particles reside on one of the branes while gravity exists in the bulk.

Our four-dimensional world resides on the ten-dimensional brane with six of the dimensions compactified. What Maldacena proposed was that our four-dimensional world full of Standard Model particles should be completely equivalent to a five-dimensional world containing quantum gravity. In Maldacena's conjecture, all the information of this five-dimensional universe is contained in our four-dimensional world. How can this be? How can you include all the information contained in our three-dimensional space in a two-dimensional sheet? At first glance, this seems impossible to implement, but there are actually examples of a similar phenomenon in modern technology. There is a good chance that you carry in your wallet a two-dimensional image that encodes three-dimensional information: the hologram on your credit card. When you look at a hologram, you can move your head around and see other sides of the object whose three-dimensional information is encoded in it.

The way that a two-dimensional hologram can contain three-dimensional information is related to the limits on the amount of information that can be stored in a given region of space or in a quantity of matter. One way to measure information content is with the concept of entropy. In thermodynamics, entropy can be thought of as a measure of the disorder in a system. More precisely, it is defined as the number of discrete microscopic states the particles that make up a system could occupy without changing the system. The entropy of an empty room in your house can be calculated by counting all the possible ways in which the molecules of air could be arranged in the room and all the ways in which they can move. If the room isn't empty, you'd need to add the same information for all the molecules that make up all the objects in the room. In a similar fashion, we can quantify the amount of information in a message by counting the number of binary digits or bits required to encode the message. This number constitutes the entropy of the message.

Physicists have applied these ideas to the study of the loss of information that occurs when matter falls into a black hole. General relativity says that once an object falls into a black hole, it will never come out again. The point of no return is called the event horizon of the black hole. However, because of the quantum mechanical fluctuations of the vacuum, pairs of virtual particles are constantly popping in and out of existence, governed by the uncertainty principle. In 1971, Yakov Zel'dovich, one of the founders of the Soviet Union's nuclear weapons program, decided to apply the ideas of quantum mechanics to the study of black holes but lacked enough knowledge of the equations of general relativity to perform the calculations. Instead of black holes requiring general relativity, he simplified the problem and used rotating metal spheres. With his crude analysis, Zel'dovich showed that the sphere should radiate energy due to its interaction with the surrounding spacetime. He published his calculations in a prestigious journal, but the paper remained mostly unnoticed.[9] The few who read it didn't think it was right.

In 1973, Stephen Hawking visited the Soviet Union and discussed the idea of the radiation of energy from the simplified black hole with Zel'dovich. Hawking didn't like the approximation that Zel'dovich had made and decided to redo it using the equations of general relativity. Hawking not only corroborated Zel'dovich's calculation that a spinning black hole would emit energy but also showed that even a nonspinning black hole would radiate energy. Hawking showed that if a pair of virtual photons appear in the curved spacetime close to a black hole, one of the two photons could fall into the black hole while the other moves away. When this happens, the two virtual photons become real. Outgoing photons created this way form a flux of thermal radiation known today as *Hawking radiation*.

Hawking maintained initially that when matter falls into a black hole, information disappears from our universe forever. He said that the radiation that comes out of the black hole is random and carries no information about what went in. The problem with this interpretation is that quantum mechanics says that information cannot be lost. If Hawking was right, quantum mechanics would be incompatible with general relativity, precluding the development of a theory of quantum gravity unless the theories were modified. Most physicists disagreed with Hawking's information loss in a black hole. In 1997, Hawking, along with Kip Thorne of Caltech, made a bet with John Preskill, also of Caltech, that if an encyclopedia fell into a black hole, all the information in that encyclopedia would be lost forever.

Hawking's calculations told him that as black holes emitted particles and radiation, they would lose mass and eventually disappear from our universe. Where would the objects and the information contained in the black hole go, if it never came back out? "The answer is that they will go off into a little baby universe of their own," said Hawking during a lecture at the University of California at Berkeley in 1988. "A small, self-contained universe branches off from our region of the universe."[10]

In 2005, Hawking redid his calculations and realized that he had been wrong.[11] Hawking radiation is related to the information that goes into the black hole. Eventually, the entire contents of the encyclopedia would come out in the radiation. He conceded his bet at a conference where he presented a paper with his new results and gave Preskill a baseball encyclopedia.

Since all the information that goes into a black hole eventually comes back out into our universe, black holes don't need to spawn out baby universes. "There is no baby universe branching off as I once thought," he wrote in his 2005 paper. "The information remains firmly in our universe. I'm sorry to disappoint science fiction fans, but if information is preserved, there is no possibility of using black holes to travel to other universes."[12]

How much information can a black hole store? To answer this question, we would need to calculate the entropy of the black hole. And how would we go about it? In 1971, Jacob Beckenstein, a Princeton University graduate student, learned that a fellow Princeton graduate student by the name of Demetrius Christodoulou and Stephen Hawking had independently discovered that the total area of a black hole horizon never decreases.[13] Beckenstein was intrigued by the similarity between this discovery and the second law of thermodynamics, which says that the entropy of a closed system never decreases. He proposed that the entropy of a black hole must be proportional to the area of the event horizon.[14] Beckenstein calculated that the entropy is actually equal to one-fourth the area of the event horizon (Fig. 9–1).

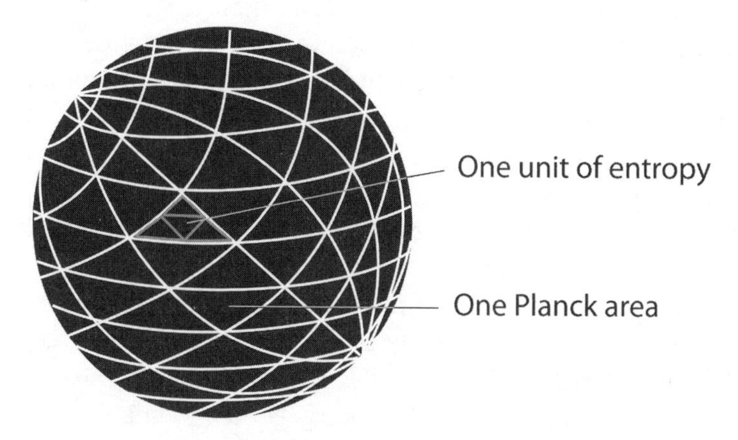

Figure 9-1. The entropy of a black hole is equal to one-fourth the area of the event horizon, the surface within which light can't escape. If the surface of the event horizon contains, say, *N* Planck areas (10^{-66} centimeter^2), represented here by the large triangles, then the total entropy of the black hole is the total area of all the small triangles.

The result seems surprising because you would expect the maximum amount of information that can be stored in a system to be proportional to the volume of the system and not to the area. In fact, when a black hole first forms, its entropy is indeed proportional to the volume. However, after the hole is formed and gravitational effects become much more dominant, the entropy scales like the area.

Beckenstein's discovery, which was corroborated later by Hawking,[15] implies that the maximum amount of information contained in a black hole depends on the area of the event horizon. And with this, Maldacena's conjecture now seemed plausible. Maldacena's conjecture is similar to a hologram in that it encodes all the information of a five-dimensional bulk into the four-dimensional spacetime, where the world of our experiences resides. The conjecture implies that there should be an alternate description of the five-dimensional world in four dimensions, much like the information contained in a three-dimensional black hole is contained in the two-dimensional area of its event horizon.

When Maldacena's conjecture is applied to negatively curved spacetimes, a consistent theory of quantum gravity emerges in which a system containing gravity can be described. The geometry of the Friedmann open universe as described in chapter 3 is an example of a negatively curved space. A space with positive curvature is called a *de Sitter space*, after the Dutch mathematician and physicist Willem de Sitter, who first studied them. A space with negative curvature is called an *anti-de Sitter space*.

According to the conjecture, M-theory in an anti-de Sitter curved eleven-dimensional spacetime is equivalent to a quantum particle theory (a gauge theory) in a flat four-dimensional spacetime. Six of the eleven dimensions in the anti-de Sitter spacetime may be compactified, as in the Hořava-Witten model. Essentially what the conjecture states is that the quantum particle theory is the hologram of M-theory. Quantum particle theories, such as the Standard Model, operating in our flat four-dimensional space, do not contain gravity but are well understood both theoretically and experimentally. M-theory, operating in the five-dimensional anti-de Sitter space with the six compactified dimensions rolled up in a tiny sphere, contains gravity but is not yet well understood. Maldacena's conjecture says that the two are equivalent. If this conjecture is true, it has enormous implications for the future understanding of M-theory.

The negatively curved spacetime is a five-dimensional anti-de Sitter spacetime with a tiny six-dimensional sphere sitting at each point. Imagine two-dimensional discs representing the five-dimensional anti-de Sitter spacetime stacked atop one another, with each disc representing the state of the universe at any one time (Fig. 9–2). This anti-de Sitter spacetime would be a three-dimensional cylinder whose axis represents time. The flat gauge theory is a quantum particle theory represented here by the boundary, the two-dimensional outer surface of the cylinder with one space dimension (the white circle) plus time. This boundary is the hologram, containing all the information of the three-dimensional anti-de Sitter space. In the four-dimensional version, which would be more like our universe, the boundary is a sphere instead of a circle.

Maldacena's conjecture tells us that M-theory and a quantum particle theory are equivalent. Being equivalent means that an entity that exists in one theory must have a counterpart in the other theory. If a particle is known to exist in one of the theories—the one describing events in the interior of the anti-de Sitter spacetime, for example—a corresponding particle or sets of particles must exist on the boundary. If one event has a certain probability of occurring in the interior, the same probability must exist on the boundary.

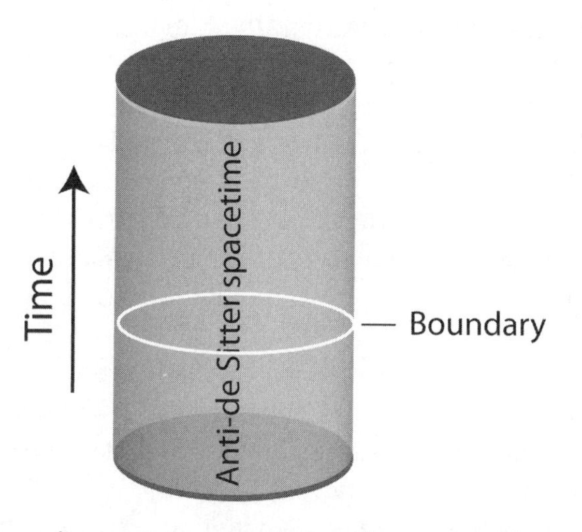

Figure 9-2. Three-dimensional representation of an anti-de Sitter spacetime. The disc is the two-dimensional anti-de Sitter space with time running along the main axis of the cylinder. The boundary has two dimensions, space (represented by the white circle) and time. The boundary is then the two-dimensional outer surface of the cylinder. The boundary is the hologram.

Maldacena found that the quantum particle theory existing on the boundary is a version of quantum chromodynamics (QCD) that includes supersymmetry. Recall that QCD is the theory that explains how quarks interact to form protons, neutrons, and all the hadrons, the particles that feel the strong nuclear force. QCD is part of the Standard Model. On the other hand, supersymmetry is an extension of the Standard Model that adds symmetry between fermions and bosons. As summarized in Table 5-2, the Standard Model contains matter particles called fermions (quarks and leptons) and force carriers called bosons. The symmetry between fermions and bosons implies that there are supersymmetric partners to all the particles of the Standard

Model, which solves one of biggest problems with the Standard Model—the hierarchy problem, the impossible cancellation of the positive and negative virtual energies.

The holographic equivalence connects M-theory, a possible theory of quantum gravity, with the supersymmetric extension of QCD. Supersymmetric QCD, which is a four-dimensional theory, is the holographic projection of the higher-dimensional M-theory. The equivalence between the two theories that the Maldacena conjecture brings provides insight into the exact equations of M-theory that still elude physicists. But the conjecture also works in the other direction. In 2005, physicists at the Relativistic Heavy Ion Collider at the Brookhaven National Laboratory were attempting to analyze a quark-gluon plasma in which quarks behave as free particles. Analyses with QCD for those extreme conditions became unmanageable, but because of Maldacena's conjecture, they were able to use techniques from M-theory that were simpler for these cases.

Although QCD has been experimentally verified many times, supersymmetric QCD—and in general, supersymmetry—has not. However, physicists are encouraged by the recent successes and think that the correspondence between M-theory and QCD should be worked out in the near future. When that happens, the efforts to develop and understand a complete theory of quantum gravity will be made more attainable. But physicists aren't waiting for the full formalism of quantum gravity to be in place. With the assumption that quantum gravity will one day be completed, they are forging ahead with their models of the quantum origin of the universe.

"CREATION OF UNIVERSES FROM NOTHING"

In 1982, Alexander Vilenkin of Tufts University published a paper that he titled "Creation of Universes from Nothing"[16] in which he proposed just that. "I would like to suggest a new cosmological scenario in which the universe is spontaneously created from literally *nothing*," he wrote in his journal paper. He reasoned that, before inflation started, when the radius of the universe was unimaginably small, the vacuum energy, which acted as attractive gravity, made the universe collapse before it could even enter the inflation phase. However, because of quantum theory, there is a small probability that the universe would not collapse, tunneling instead through the potential barrier to a larger radius where it started inflating. Vilenkin then calculated the quantum mechanical tunneling probability for universes of smaller and smaller radii and discovered that the probability remained finite when the radius of the universe approached zero. In the limiting case, when the radius was exactly zero, there was a finite probability that the universe could tunnel from nothing into the inflationary phase.

In Vilenkin's calculations, the universe popped into existence from nothing and entered the inflation phase, which was followed by the standard big bang and the rest of the evolutionary history of the universe. But, if the universe suddenly popped out of nothing, what caused it to appear in the first place? According to quantum mechanics, the tunneling process is random and unpredictable. For example, when a nucleus undergoes radioactive decay, it does so at any time without any known cause. Although the half-life of carbon-12 is 5,730 years, we can't predict exactly when a given nucleus of carbon-12 will decay. It can decay during the next second or in the next 100,000 years. A macroscopic sample of carbon-12 contains an enormous number of nuclei and the laws of probability can tell us how many nuclei—but not which ones—will decay in a certain length of time. In 5,730 years, half of the carbon-12 in our sample will decay. In another 5,730 years, half of the remaining half will decay so that after 11,460 years, a quarter of the initial amount remains. In 17,190 years, or three half-lives, an eighth of the initial number of nuclei remains. Since we can measure this half-life very accurately, scientists can use radioactive carbon-12, which is taken in by all living organisms, to accurately date archeological artifacts made from organic materials. Like radioactive decay, the tunneling of the universe from nothing is controlled by quantum mechanics and is random, requiring no particular cause.

Vilenkin used an elegant semiclassical description of quantum tunneling that had been suggested by Sidney Coleman of Harvard University.[17] Coleman suggested using *Euclidean time,* a mathematical tool used in quantum mechanical calculations in which the time coordinate is expressed in imaginary numbers. The advantage of imaginary time is that it behaves just like one of the spatial dimensions, increasing or decreasing in value as the spatial direction does. With Euclidean time, instead of a four-dimensional spacetime with three dimensions of space and a distinct time dimension, we have a four-dimensional space. Since they are all space dimensions, they all can have a shape.

With Euclidean time in the equations, spacetime can be finite with no sigularities. A four-dimensional spacetime has a singularity at time zero (Fig. 9–3 Left) where the equations of general relativity break down. In Euclidean time, the equations describe a four-dimensional sphere that eliminates the singularity (Fig. 9–3 Right). Vilenken interpreted his solution as describing the tunneling from nothing so that the origin of the universe could be symbolically represented as shown on the right side of Fig. 9–3.

But what really is "nothing"? For Vilenkin, "nothing" is a universe with zero radius, which is the same as no universe. However, his "nothing" still obeys the laws of physics. According to his idea, the laws of physics existed before the universe came into being. Although Vilenkin's scenario provided a cosmological model without a singularity at the big bang, the creation from nothing required the preexistence of the laws of physics.

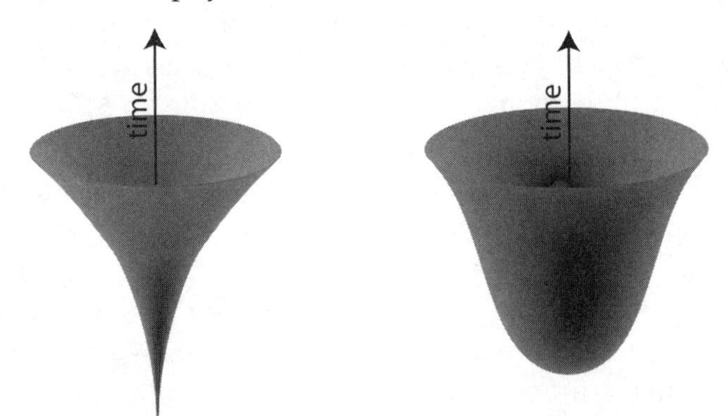

Figure 9-3. The diagram on the left illustrates the evolution of the early universe in a four-dimensional spacetime with a singularity at the big bang. On the right, the solution with Euclidean time eliminates the singularity. Spacetime is finite but has no boundary. This scenario describes a universe tunneling into existence from nothing.

A Universe with No Boundaries

A year after Vilenkin's paper was published, Stephen Hawking, working with James Hartle of the University of California at Santa Barbara, published a paper stating that the universe has no boundaries in space or time.[18] While this idea appears to be the same as what Vilenkin had already said, Hawking and Hartle's approach was different and more general. What they did was to discover a way to calculate the quantum state of the entire universe, which is a way to calculate the probability that the universe is in a certain state. To calculate this quantum probability they had to include all the possible histories of the universe. In their approach, a version of *Euclidean quantum gravity*, this superposition of all the possible histories is done in Euclidean time.

The idea of multiple histories is a quantum mechanical concept invented by Richard Feynman of Caltech in 1965. Feynman, who was one of the most creative physicists of his generation, was famous for inventing very powerful methods that are easy to understand and use and substituting them for many complicated mathematical techniques. One of those methods was the *sum over histories*, also known as the *path integral approach*. To see how Feynman's method works, consider again the experiment with electrons passing through two narrow slits as described in chapter 5. When no detectors are placed behind each slit to watch the electrons and see through which slit they pass, the electrons strike the screen at strange locations. After many electrons pass through the two slits, the points where they strike the screen form a pattern of parallel bands called an interference pattern (Fig. 9–4).

As I explained in chapter 5, when we observe each electron as it passes through either slit, the electrons hit the screen just as particles would: most hits are directly in front of the slits, with some scattered hits from electrons that get slightly deflected by the edges of the slits. When we don't observe the electrons as they cross the slits, they form an interference pattern. Could it be that the electrons collide and somehow interact with each other in such a way that they end up on the screen forming a pattern similar to an interference pattern? The possibility of something like that is extremely remote, but we can still examine it. We could reduce the intensity of the electron beam until we have just one single electron emitted at a time. At first, we see flashes on the screen at what appear to be random locations, without seemingly forming any pattern. When enough hits are recorded on the screen, however, the interference pattern reappears. A single electron forming an interference pattern, interfering with itself!

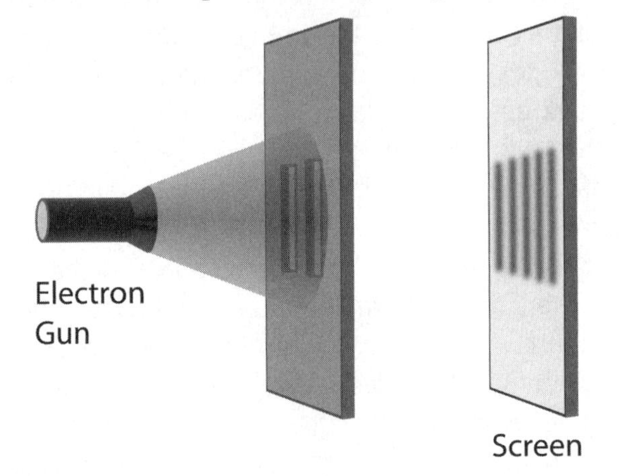

Figure 9-4. Electrons from an electron gun pass through two parallel slits and strike a screen behind. When the electrons are not observed, the strikes on the screen form an interference pattern.

The single electron interference takes care of the interaction between the electrons but brings up the strange phenomenon of a single electron interfering with itself. One answer to this mystery is the one I've already given: when we don't observe the electron, it behaves like a wave and produces interference, a wave phenomenon. But Feynman had another answer: the electron passes through both slits at once. In the language of quantum mechanics, we say that when we don't observe it, the electron is in a superposition of the state in which it passes through one slit and the state in which it passes through the other slit.

In Feynman's sum over histories approach, when an electron travels from point *A* to point *B*, it doesn't simply follow one path. Instead, the electron travels along every possible path through space-time (Fig. 9-5). One particular path might take the electron from the gun to the slits via the other side of the Earth or to the moon, for that matter. The probability that an electron travels from *A* to *B* is the sum of all the contributions from all the possible paths between *A* and *B*. Now, a baseball travels from the pitcher's hand to the catcher's mitt following a single path. The reason is that, for macroscopic objects, when you add all the contributions from all possible paths using Feynman's method, contributions from similar paths cancel each other out and only one is left. The contribution that remains is the one given by Newton's laws of motion.

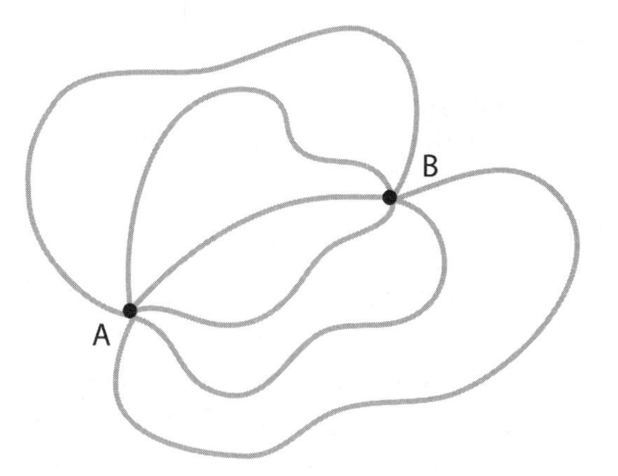

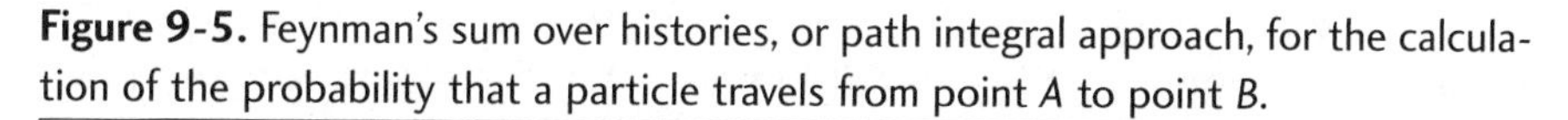

Figure 9-5. Feynman's sum over histories, or path integral approach, for the calculation of the probability that a particle travels from point *A* to point *B*.

The calculations of the sum over histories (or path integral) approach are done in the Euclidean four-dimensional space, with four spatial dimensions instead of three spatial dimensions plus time.

Hawking and Hartle decided to apply this method to their calculation of the quantum state of the universe; that is, to compute the probability that the universe is in the state that we observe it to be. To perform that calculation in quantum mechanics using Feynman's method, they needed to include all the possible histories that the universe followed to get to the present state. When performing such a calculation for a subatomic particle, physicists select the fundamental events that make up the valid histories to be included. Consider, for example, the case of an electron moving in a region with a barrier (Fig. 9–6). We all know that if instead of an electron, we have a tennis ball thrown at a lower height than the net, the probability that the ball crosses the net is zero and the probability that it bounces back is one. Tennis matches would be much more exciting if this weren't the case!

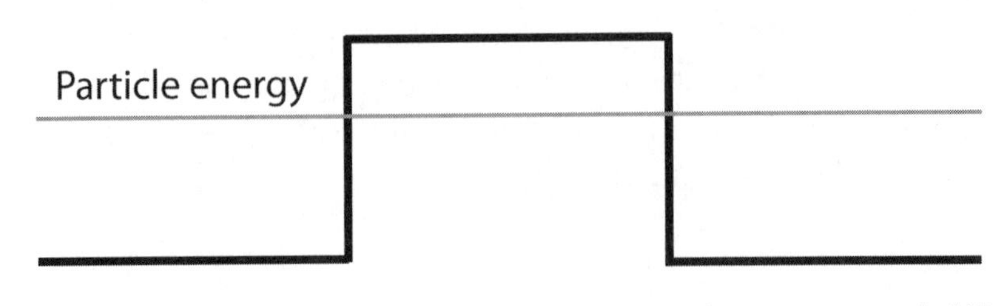

Figure 9-6. An electron moving in a region with a barrier has a certain probability of tunneling through the barrier as well as a certain probability of bouncing back.

However, in the case of an electron, the situation is different. Quantum mechanics predicts that there is a certain probability that the electron will tunnel through the barrier and appear at the other side, but there's also a certain probability that the electron will bounce back. Since these are the only two possible options, the sum of these two probabilities must equal 100 percent. To calculate the probability that the electron is found in the state in which it tunnels through the barrier, we need to include all the histories of the electron before it got to that state; that is, all the possible paths that the electron could have taken to get there. The left side of Fig. 9–7 shows three of the possible histories. The first history contains three processes: transmission across the first wall, propagation through the barrier, and transmission across the second wall. The second history contains five processes: transmission across the first wall, propagation through the barrier, reflection from the second wall, propagation again through the barrier, followed by reflection at the first wall, propagation once more, and finally transmission through the second wall. The figure shows a third history, but there is an infinite number of histories. Each history contributes to the overall value of the probability, but the first terms contribute the most and physicists find that a few histories are all that is needed to get the right answer. A similar process takes place in the calculation of the probability that the electron is found in the reflected state, as seen on the right of Fig. 9–7.

For the calculation of the probability that the universe is in the state that we observe it to be, we need to determine the possible histories. The choice of which histories to include affects the calculation of the overall probability. Hawking and Hartle chose to include the simplest class of histories, which is a four-dimensional sphere in the Euclidean space that the sum over histories is done (Fig. 9–8). Closed surfaces have no boundaries. As Hawking is fond of saying, "The boundary condition of the universe is that it has no boundary."[19]

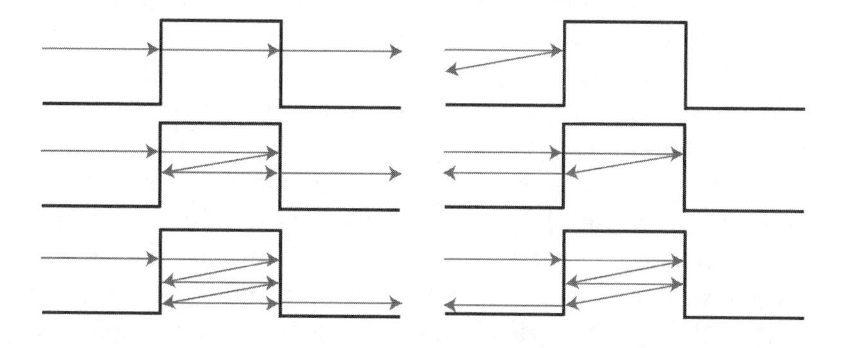

Figure 9-7. Histories for the transmission and reflection events of an electron moving in the presence of a barrier.

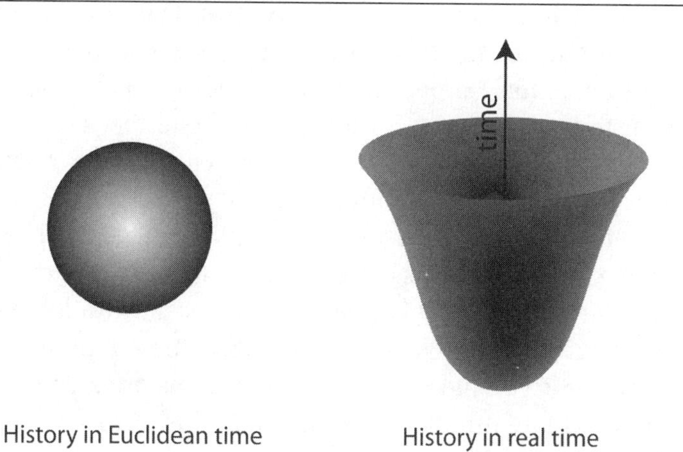

Figure 9-8. The simplest history of the universe is a four-dimensional sphere in Euclidean time. This Euclidean four-sphere corresponds to an expanding universe in real time.

The closed Euclidean four-sphere corresponds to an inflating universe in real time, but this is a universe with an inflationary era that will not stop. This history also leads to a universe without stars or galaxies, which is not the universe that we observe. However, if the Euclidean four-sphere is not perfectly smooth and slightly flattened at the south pole, the real time universe will initially inflate, but the inflation will slow down and turn into a big bang expansion. The tiny bumps and dimples of the imperfect Euclidean sphere will give rise to stars and galaxies and to structures similar to what we observe in the universe today.

Since the possible histories of the universe are infinite, how do we make sure that we have selected the right set? In the simple case of the electron moving in the region with a barrier, the boundary conditions set by the barrier are well defined and the choices are clear; the only decision is where to stop including similar histories. In the case of the entire universe, the situation isn't clear by any stretch of the imagination. Most histories will produce a universe that could not harbor galaxies, stars, and life, and that isn't the universe we see. Even when the choice of histories is restricted to Euclidian spheres, there won't be one single history but a whole family of histories. According to the uncertainty principle, many spheres with slightly different deformations will lead to possible universes. Only in those in which the deformations are small will stars and galaxies evolve.

For Hawking and Hartle, the universe we see determines the set of possible histories. Our set of histories is the one that led to intelligent beings who can observe the universe and develop theories about its existence. This is, of course, the anthropic principle. The history of the universe led to the universe we see because if it hadn't, we wouldn't be here.

Hawking and Hartle assumed that a few simple contributions dominate the sum over histories. More recent results with other quantum gravity approaches indicate that this simplified approach might not be correct and that the sum over histories could be dominated by an infinite number of complicated histories.

PARALLEL UNIVERSES

When we apply the quantum mechanical wave equation to a system such as an electron moving in a region with a barrier or in the vicinity of an atom or even to the entire universe, the answer that we get is not a fixed number but a probability. If we use Newton's physics to describe the path of a ball that we throw at a six-foot-high wall, the answer we get is not a probability but a certainty. The answer we get is that if the ball travels in a horizontal path one foot off the ground, for example, the ball will bounce back and move in the opposite direction. If we know the speed at which we throw the ball and its elasticity and if we take into account the drag force as the ball moves through the air, we can get a very accurate value of the speed at which the ball bounces back. We can even make the calculation before we throw the ball and, if we are skillful enough to throw it exactly at the height and with the speed that we used in the calculation, the ball will bounce back exactly as the calculation says it would. NASA has been launching spacecraft to the planets for decades using only Newton's physics, reaching Mars, Jupiter, and the outer planets as the calculations predicted.

But if an electron is traveling with an energy of one unit in a region where there is a barrier that has a height of two units and we want to know the motion of the electron as it interacts with the barrier, the answer we get back is that there is a certain probability that the electron will bounce back and a certain value of the probability that the electron will tunnel through the barrier and move to the other side. But it is impossible to predict exactly whether the electron will bounce back or will tunnel through before it actually happens.

The strict quantum mechanical description of the state of an electron is done with a wave function—a solution to the equation of motion of a quantum system that is known as the *Schrödinger equation*. This equation precisely describes the evolution of an electron or of any quantum system. However, moving from the elegant mathematical description of the Schrödinger equation to the macroscopic world of our experiences has proven to be difficult. Is the quantum world, with its seemingly strange behavior, separate and distinct from our macroscopic world? Or are we quantum mechanical systems along with the rest of the macroscopic universe?

The orthodox interpretation of quantum mechanics, developed by Niels Bohr and many of the leading physicists of his time shortly after the discovery of the theory in the 1920s, takes the first approach. The quantum world and the world of our experiences are different. The observer of a quantum event exists in a classical world that is different and separate from the quantum world of the subatomic system being observed.

Bohr's interpretation became known as the *Copenhagen interpretation*. According to Bohr, an electron moving in a region with a barrier exists in a superposition of two states, one in which the electron bounces back from the barrier and the other one in which it tunnels through the barrier. Similarly, the electron in the two-slit experiment as described earlier exists in a superposition of two states, one in which the electron passes through one of the slits and the state in which it passes through the other. The electron does not have a position until you look. It actually acquires a position at the moment of the observation, when the wave function *collapses* into one of the alternatives. The question of the actual location of the electron is therefore meaningless without an observation to locate it. In Bohr's interpretation, the wave function collapses from the superposition of two states to the actual state in which it is found when the observation is made. At that moment, all but one of the alternative states vanish. Thus, in the Copenhagen interpretation, the act of observation interrupts the smooth evolution of the wave function. Moreover, this collapse of the function is not part of the theory but an added postulate.

The second interpretation, proposed in 1957 by Hugh Everett III, a Princeton graduate student working under John Wheeler, takes the second approach. For Everett, the microscopic and the macroscopic world are both in the quantum realm. There is no separation between the observer and the subatomic system being observed. Since everything is described by quantum mechanics, there is no break in the description of the wave function when an observation is made.

Instead of imposing a postulate on the theory, Everett let the theory guide him, taking quantum mechanics at face value. Instead of a collapse of a wave function into only one of the possible states that made up the superposition, Everett's wave function continues to exist in all the possible versions that correspond to the possible outcomes. In the case of the wave function of an electron in a region with a barrier, for example, once the electron interacts with the barrier, the universe branches out in two copies of itself, one containing an electron that bounces back and the other with an electron that tunnels through. These two parallel universes do not interact with each other.

In Everett's view, instead of a wave function that describes the electron for a classical and independent observer to observe, there is a universal wave function that describes the system and the observer. If you are observing this electron, you and the electron split into two parallel universes, one that contains you and the electron that bounced back and the other with you and the electron that tunneled through.

Every interaction between systems generates branches of the universe, each containing one of the possible outcomes. If Everett's interpretation is correct, three seconds ago you were reading the words in the preceding sentence. In a parallel universe, three seconds ago you started reading the previous sentence but didn't finish because the phone rang. At that moment, an exact copy of you in that parallel universe stopped reading, stood up, answered the phone, and were told to rush to a friend's house because of an emergency. That same moment, another exact copy of you in yet another parallel universe decided that it was time to have coffee and stood up to brew it.

Do you find this ridiculous? If you do, the relatively few physicists who at the time read Everett's original paper would have agreed with you.[20] John Archibald Wheeler, the renowned Princeton professor who was Everett's PhD thesis advisor was, on the other hand, very enthusiastic about his idea. Everett had developed his idea into an alternative and more consistent interpretation of quantum mechanics and wrote it up as his doctoral dissertation.[21] The thesis was rigorous and dealt strictly with the formalism of quantum mechanics. Seeing how revolutionary this work was, Wheeler decided to show it to Bohr and traveled to Copenhagen exclusively to discuss Everett's thesis with the great physicist. Bohr wasn't convinced.

Unwilling to challenge Bohr, Wheeler told Everett to rewrite his thesis, toning down its conclusions.[22] Everett reluctantly complied, reworking his thesis and reducing it to about a quarter of its original length.[23] By the time his thesis committee approved the thesis, Everett had already taken a job with the Pentagon to do mathematical analyses of the effects of radioactive fallout from a nuclear war, eventually advising presidents Eisenhower and Kennedy on the optimal selection of nuclear bomb targets. Disillusioned with quantum mechanics because of the criticisms and the general lack of interest in what he thought was of outmost importance for theoretical physics, Everett continued working for the Pentagon after formally obtaining his doctorate. He would never return to quantum mechanics.

The *many worlds interpretation* of quantum mechanics, as Everett's interpretation became known, started to be noticed in 1967, when Bryce DeWitt of the University of Texas, who initially opposed his view, credited Everett for showing the need for the Wheeler-DeWitt equation. DeWitt and his graduate student Neill Graham later edited a book of contributions on the topic by other researchers and featured the original longer version of Everett's thesis.[24]

The current view of the many worlds interpretation is that there is only one quantum multiverse with one wave function that evolves deterministically, without undergoing splitting of any kind. As Max Tegmark of the University of Pennsylvania writes, "The abstract quantum world described by this evolving wave function contains within it a vast number of parallel classical story lines, continuously splitting and merging."[25] A process known as *decoherence* prevents observers from seeing the other worlds containing their copies. These observers perceive only the partial reality of their own universes. Each observer sees what appears to be a wave function collapsing into the observed reality of her or his world while all the other possibilities instantaneously disappear. The discontinuity in the evolution of the Schrödinger equation, which manifests itself as randomness in the quantum world, is due to our inability to observe the parallel realities.

Although decoherence can be used to describe the many worlds interpretation, the two are not the same. Decoherence is the process by which obtaining information about a system destroys the wave nature of that system.[26] Thus, decoherence provides a simpler alternative to the wave function collapse explanation of the problem of measurement in quantum mechanics, but it does not imply that the measurement problem splits the world into multiple copies. The measurement problem arises from the components of the wave function—the histories—that don't actually happen. What decoherence does is to allow you to ignore these other histories, these other parts of the wave function in the cases in which they don't have any influence on the outcome that is observed. Since these histories can be ignored as soon as you can show that they don't influence our world, decoherence depends on the present as well as on the future behavior of the system.

In the case of the two slit-experiment, there are only two histories corresponding to each one of the slits through which the electron can pass. These two histories interact with each other to produce the interference pattern on the screen. These histories affect the final outcome and cannot be discarded: they are coherent.

If we now add detectors behind the slits to monitor the passage of the electrons, we still have two histories, but now we know when the electron crosses the slits and the interference pattern disappears from the screen. The presence of the detectors, that is, our ability to observe the electrons as they pass through the slits, destroys the interference. Since these two histories do not interact with each other to produce interference, their wave nature has been destroyed and they are said to be decoherent. Thus, measurement destroys coherence.

The decoherent histories interpretation of quantum mechanics provides a satisfactory explanation of the measuring problem. We still have the superposition of states, but instead of a collapse of the wave function at the moment of observation, when all the histories but one disappear, decoherence allows us to keep all the histories. One becomes reality. The others are still there but we ignore them because they do not affect the future history of the system. And they don't necessarily create other parallel universes.

As we have just seen, the solutions offered to the problem of origins are still incomplete, although there are several promising possibilities. The success of the inflationary theory has brought to the front burner the need for cosmological initial conditions. Searching for those conditions appears to be possible only through quantum theory, and this requirement underscores the need for a full understanding of the foundations of quantum mechanics. The decoherent histories approach and the many worlds interpretation offer solutions to this problem. Recent work using a decoherent histories quantization of simple quantum cosmological models, although in its early stages, is showing the way toward a better understanding of the conditions of the preinflationary early universe and to the full solution of the problem of origins.

Chapter 10

THE SELF-SELECTING UNIVERSE

THE NO BOUNDARY PROPOSAL

We think that we have a good idea about how the early universe rapidly evolved and gave rise to the minute thermal fluctuations that one day would turn into stars, galaxies, and all the structure that we see around us. But the biggest question of all still seems to elude us: the question of how it all began. There are possibilities, to be sure. Among them is James Hartle and Stephen Hawking's no boundary proposal, which provides an innovative new answer to that question: the universe has no origin.

In their original proposal, Hartle and Hawking assumed that, since quantum gravity was essential for the early universe, they could apply the quantum sum over histories method to calculate the quantum state of the universe. They recognized early on that the possibility of finding the present state of the universe in most of the possible histories was negligibly small and decided instead to sum over a simple class of histories that is more likely to describe our universe. The simplest form of this class of histories is the four-dimensional Euclidian sphere. In this Euclidean spacetime, there are no boundaries, no singularities.

However, further work by other researchers showed that Hartle and Hawking's original proposal was incomplete in several different ways. Don Page of the University of Alberta identified a number of problems.[1] The first problem is that in quantum gravity, the path integral or sum over histories is ultraviolet divergent and not renormalizable—which, in plain English, means that the integrals give infinite answers at high frequencies and the infinites cannot be removed.[2] Although this wasn't a problem that arose from the model itself but from quantum gravity, the model is based on the path integral method.

Another problem is that in the sum over histories, we must include all four-dimensional topologies. Topology is the study of the properties that remain through any deformations of mathematical objects. For example, a circle and an ellipse have the same topology since the circle can become an ellipse by stretching it. The problem for the Hartle-Hawking proposal is that there is no method to decide whether two four-dimensional manifolds have the same topology. This problem doesn't arise from the proposal either and is more pertinent to M-theory, where it might one day be solved.

A more relevant problem is that even if the histories are uniquely defined, the calculation of the integrals for all the paths is extremely difficult. It is likely that only approximate solutions can be obtained, which would obviously provide us with approximate descriptions of the universe.

DELAYED CHOICE EXPERIMENT

In 2006, Hawking, working with Thomas Hertog of CERN, proposed a new version of the Hartle-Hawking no boundary proposal.[3] The new proposal combines the M-theory landscape with the no boundary initial conditions, resulting in what they call a top-down approach to cosmology. In their approach, the histories of the universe depend on what we observe today, "on the precise question asked," as Hawking and Hertog write. Our observations of today select the entire history of the universe that is consistent with our existence.

The observer-selected reality—an aspect of the quantum world—has been seen in the laboratory. To see what led to such an observation, consider once more the double-slit experiment with an electron gun shooting electrons at two parallel narrow slits. As described in chapter 9, when we monitor the passage of the electrons through the slits with small detectors, the electrons behave as we expect particles to behave. They hit a screen placed behind the slits mostly at locations behind the slits, although some are deflected by collisions with the slit edges and hit farther away. Nothing mysterious here. However, if we remove the detectors and don't monitor the passage of the electrons through the slits, an interference pattern promptly appears on the screen, telling us that the electrons are waves that follow the two paths simultaneously.

When we don't attempt to measure the position of the electron to determine through which slit it passes, it behaves like a wave and produces interference. When we do look, it behaves like a particle. From our point of view, the obvious question is: How does the electron know what we are doing? We could argue that to be able to monitor the electron passage through the slits, the detector has to interact with the electron, disturbing it. According to the Heisenberg uncertainty principle, attempting to determine the electron's position increases the uncertainty in its velocity and that, in turn, destroys the interference pattern.

But here's the kicker. We can delay the choice of whether to look for particles or for waves until *after* the particle has passed through the slits but before it hits the screen. What would happen then?

In 1978, John Wheeler proposed such an experiment.[4] His *delayed choice* experiment was a *Gedanken*, or thought experiment, since at the time he didn't think it was possible to perform it in a laboratory. Technological advances since then have made the real experiment possible and to date, no fewer than seven Wheeler experiments have been performed.[5] The most recent experiment, and up to now the only one that closely matches Wheeler's proposal, was performed in 2006 by Vincent Jacques and his collaborators in France.[6]

As in Wheeler's idealized arrangement, in Jacques's experiment a single photon is sent toward a beam splitter (BS_{in}) that generates two possible paths and that plays the role of the two slits in the two-slit experiment (Fig. 10–1). In this single photon Mach-Zehnder interferometer,[7] as the arrangement is known, two mirrors deflect the beams toward two detectors 1 and 2. A second beam splitter (BS_{out}) can be introduced or removed at will. The choice of inserting or removing the second beam splitter is decided at random using a device called a Quantum Random Number Generator. To provide enough time to achieve the delayed choice of inserting or removing the output beam splitter, the interferometer was designed with a length of 48 meters (157 feet).

In its *closed* interferometer configuration, with BS_{out} in place near the detectors, a single-photon pulse encounters the first beam splitter, BS_{in}, and travels through the interferometer until the second beam splitter recombines the interfering paths. In this case, an interference pattern is detected—"evidence," as Wheeler wrote, "that each arriving light quantum has arrived by both routes."[8] In its *open* configuration, when the second beam splitter is removed, each detector is clearly associated with one single path. If the photon travels along path 1, it enters detector 1, which registers the event. If, on the other hand, the photon travels through path 2, it enters detector 2, which registers that event. In the closed configuration, then, either "one counter goes off or the other: Thus the photon has traveled only one route."[9]

Jacques and collaborators used an electro-optical device with a specially designed fast driver to switch the second beam splitter in and out of the interferometer. The fast driver was able to switch the splitter in 40 nanoseconds. Since the photon takes 160 nanoseconds to traverse the 48-meter distance between the locations of the two beam splitters, there was plenty of time to switch between the open and closed configurations. The decision to place or not place the second detector—clearly too fast for a human—was made by the Quantum Random Number Generator.

The delayed-choice experiment was performed with the electro-optical device randomly activating the switching of the beam splitter for each photon sent into the interferometer. In the closed configuration, with the second beam splitter placed in the interferometer *after* the photon had passed the first beam splitter, the scientists observed an interference pattern, characteristic of the wave behavior of the photon. In the open configuration, when the electro-optical device randomly removed the second beam splitter from the interferometer, the team observed that the interference pattern disappeared completely. The photons, passing one by one through the interferometer during the times that the second beam splitter was not in place, all behaved like particles.

Jacques's team showed experimentally that the behavior of the photon as particle or as wave depends on the choice of what we measure. When we set out to look for electrons as particles, we measure particles, and when we set out to measure the wave nature of the electron, we observe its wave behavior. What is really crucial in this remarkable experiment is that the choice of what property we measure is done after the photon has passed through the splitter and is either traveling through both paths at once as a wave or through one of the two paths as a particle. We can no longer say that the presence of the detectors interacting with the photons somehow destroys the interference pattern because the decision was made after the photon passed the splitter. Whether the photon shall have traveled across the interferometer by one path or both is decided *after* the photon has started its trip. As Wheeler wrote, "We have the strange inversion of the normal order to time. We, now, by moving the [beam splitter] in or out have an unavoidable effect on what we have a right to say about the already past history of that photon."[10]

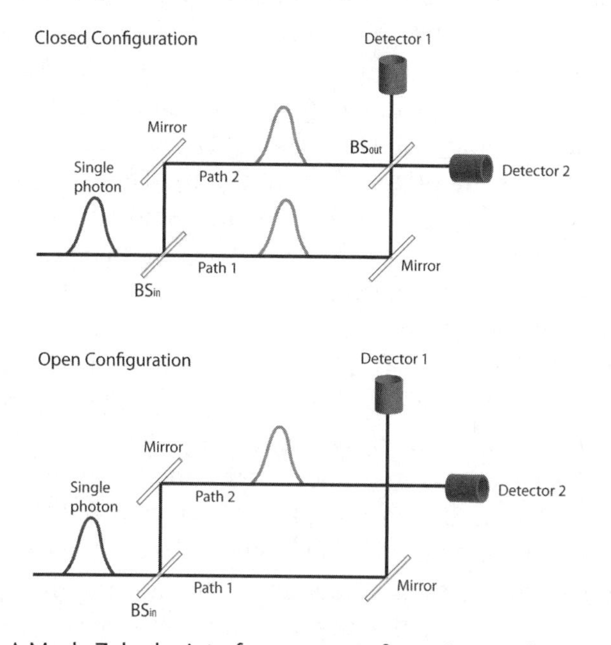

Figure 10-1. A Mach-Zehnder interferometer configuration used in the implementation of Wheeler's delayed choice experiment. A single approaching photon first encounters a beam splitter (BS_{in}) that generates two possible paths. In the closed configuration (top), a second beam splitter (BS_{out}) is placed near two detectors located at the end of the two possible paths. In this configuration, the photon, traveling along both paths at once, generates an interference pattern, characteristic of waves. In the open configuration (bottom), the second beam splitter is removed and the photon behaves like a particle traveling along a single path, either path 1 or path 2. The decision to insert or remove the second beam splitter is made *after* the photon has passed through the first beam splitter.

A TOP-DOWN APPROACH TO COSMOLOGY

Hawking and Hertog propose that, just as our decision to move the beam splitter in and out of the interferometer selects the past history of the photon, our present observations select the past histories of the universe from among all the possible histories. In this new version of a Euclidean quantum gravity called the *top-down* approach, we trace the histories of the universe backward. Their proposal is a refinement of the original no boundary proposal of Hartle and Hawking.

In the 1960s, Hawking and Roger Penrose proved the now famous singularity theorems of general relativity that provided the conditions for the formation of black holes. A consequence of those theorems is that the equations of general relativity imply that a classical cosmology has a singularity at the origin of the universe. Therefore, general relativity, a classical theory, breaks down at this point and cannot be used to study the origin of the universe. Hawking interprets the theorems to imply that the origin of the universe was a quantum event and, as such, should be described by the sum over histories method of quantum mechanics. When applied to a quantum system such as an electron traveling between two points, physicists need two boundaries: the initial state and the final state. If the electron emitted from a cathode (the coated tungsten filament) in a fluorescent tube is deflected by a negatively ionized atom in the plasma of the tube, we would need to know the energy with which the electron is emitted from the cathode (initial boundary) and the physical properties of the ion (the final boundary). The ion acts as a barrier to the motion of the electron, so we would need to know the height or potential of the barrier as well as its width. With the two boundaries, we could begin to include the contributions from all the possible paths—the histories—between the two points.

But in the case of the universe, we don't know the initial state and, according to Hawking, we probably never will. That isn't a problem for Hawking and Hertog because, in their view, the quantum state of the universe at the present time is independent of the initial state. The reason is that there are cases in Euclidean time where the initial and final states are in separate and disconnected regions. The final state may exist in one universe and the initial state might have existed in another. Since we don't know the initial state, the final state would be the sum over all histories whose single boundary is the present quantum state of the universe. This is the no boundary quantum state.

The new view leads to a radical new path in cosmology that changes the relation between cause and effect. Since the only boundary that we can use is the present quantum state of the universe, we must follow the histories of the universe backward in time. Following the histories backward in time is possible in Euclidean space, where space and time are on an equal footing and where we simply have four spatial dimensions and no time dimension. Therefore, the histories that we must include in the sum over histories method are not independent—they depend on the probabilities that are measured at the present time. The histories depend on what we observe today. In the no boundary proposal, there is a finite probability for universes with zero, one, two, all the way to ten spatial dimensions, and there is one history of the universe for each one of those possibilities. There is a history of the universe in two spacetime dimensions, for example, or one in nine large spacetime dimensions, or in eleven. But we observe only four spacetime dimensions in our universe today. In the top-down approach, then, it doesn't matter how large the probability is for a universe with seven large spacetime dimensions because that's not the universe we observe. As long as the probability for a history of the universe with four spatial dimensions is not zero, that is what we include in our sum over histories calculation. We do not include any of the others, regardless of how large the probability of their existence, because the universe that we observe doesn't have them. In his often-colorful metaphors, Hawking compares this aspect of the top-down proposal with the probability of his being Chinese. He knows he's British, even if the probability of a human being British is much smaller than that of being Chinese. Since the probability of being British isn't zero (we know that there are some British in the world!), he's safe.

What we observe and what we measure in our universe today determines the histories that we include in the sum over histories that gives us the wave function of the universe. Therefore, in addition to the history of the universe with four large spacetime dimensions, we would include a history for a universe with the observed cosmological constant, a history of the universe in which the gravitational interaction has the measured strength today, a history with the measured strength of the weak nuclear force, and so on.

If the probabilities for histories that the top-down approach does not include aren't zero, where are those other universes? Hawking and Hertog think that these are the possible universes in the landscape of multiple stable and metastable vacuum configurations of M-theory. These alternative histories might have existed for a brief time during the first instants after the big bang. During this epoch, the very early universe was in a superposition of all these quantum states. Most of those alternative histories disappeared, leaving the history that led to our universe today. But those histories left a trace in our own universe. "Our current universe has features frozen in from this early quantum mixture."[11]

Since our universe has the early alternative histories imprinted in it, we should in principle be able to detect these imprints in the cosmic microwave background radiation. The proposal would then be testable. As described in chapter 4, these fluctuations were measured with unprecedented precision by NASA's Wilkinson Microwave Anisotropy Probe (WMAP) in 2006 (Plate IX). Hawking and Hertog are refining their calculations with their model and hope to obtain their own spectrum of fluctuations to compare with WMAP's observations.

EXPLAINING THE FINE-TUNING

For Hawking, explaining the remnants of the big bang, the fluctuations that generated the structure of the universe, constitutes the central problem of cosmology. Why is the universe the way it is? Why does the cosmological constant have the value that it has, a value that allows for the emergence of life?

As described in the preceding chapters, in the traditional approaches to cosmology, there isn't yet a clear, proven mechanism to determine the value of the cosmological constant for our universe—a value so small that it cancels out to 120 decimal places—although there are promising attempts, such as the cyclic model. However, Hawking and Hertog argue that the only way to explain the fine-tuning observed in our ordered universe today is not to assume an observer-independent single history of the universe, but to consider the probabilities for the possible histories of the universe using boundary conditions at later times; that is, histories traced backward in time, as in their top-down Euclidean quantum gravity approach.

Since the probabilities for the histories of the universe depend on what we measure today, the fine-tuned value of the cosmological constant that we calculate from our observations constitutes one of the boundary conditions that constrain histories that we include in the sum over histories. Likewise, the precise equilibrium of the laws of physics that we observe today represents a boundary condition that fixes that particular history. Hawking and Hertog's proposal doesn't need to explain the mechanism that gave rise to the laws of physics or to the cosmological constant. They actually think that it is really not possible to ever know the exact details of how our universe got going at the big bang.

Hawking and Hertog think that if their top-down Euclidean quantum gravity proposal is correct, the central problem of cosmology is solved. The universe is the way it is, the laws of physics are what we observe, and the cosmological constant has the precise value that it has because those are the parameters and laws that we observe today. These observations are the final boundary conditions that we use to calculate the probabilities for histories of the universe with those exact parameters. If those probabilities turn out to have a nonzero value in the calculations of their model, those are the histories that we include in the sum over histories. And those probabilities need not have the largest values among the possible probabilities.

However, the top-down approach does not explain why the laws of physics are what we observe or the value of the cosmological constant is what we measure. The top-down approach doesn't depend on that particular value of the cosmological constant. But this value is in the very narrow range that allows for the existence of life in the universe, as discussed in chapter 8. Our universe has this specific value of the cosmological constant and we chose that value as the final boundary for one of the probabilities that we include. Similarly, we select four macroscopic dimensions as a final boundary because that is what we observe in our universe.

Hawking emphasizes that although the selection of four macroscopic dimensions as a final boundary may appear to be based on the anthropic argument because life seems to be possible in a universe with four dimensions, the selection is actually based on a different argument. The selection of four dimensions as a final boundary or of the specific value of the cosmological constant does not depend on whether four dimensions is the only possibility for life or our value of the cosmological constant is the only possibility for the emergence of life. "Rather, it is that the probability distribution over dimensions is irrelevant because we have already measured that we are in four dimensions."[12] We don't select four large dimensions as a final boundary or the measured value of the cosmological constant as another final boundary because life exists in four dimensions or in universes with this value of the cosmological constant, but because those are our present observations. The selection says nothing about the possibility of life in five large dimensions or in other universes with slightly different values of the cosmological constant.

AN INFLATIONARY NO BOUNDARY UNIVERSE

Hawking and Hertog have constructed a simple model using their top-down approach as a first approximation. Their simplified model is described by the evolution of a field moving in a potential with the shape of a Mexican hat (Fig. 10–2). This potential has a wide, flat central maximum. The minimum value is the small value of the observed cosmological constant. They used their simplified model to calculate the no boundary probability of an expanding region of spacetime that is similar to our universe.

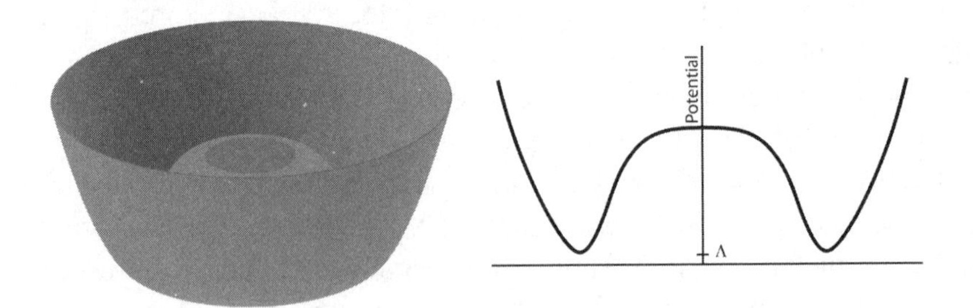

Figure 10-2. The Mexican hat potential with a wide flat central maximum and a minimum value equal to the value of the cosmological constant used by Hawking and Hertog in their simplified top-down model of the universe.

Hawking and Hertog calculated the wave function in Euclidean space, which allowed them to use their model to compute the probability of an inflating universe. The field starts in the middle of the plateau, the wide, flat region of maximum potential at the center of the hat. Quantum fluctuations cause the field to roll down the potential hill until it reaches the bottom of the hat, where it oscillates around the minimum value until it settles at the currently observed value of the cosmological constant. This history, determined by the final boundary at the present, has a relatively long period of inflation and produces a perfectly flat universe containing stars and galaxies, like our universe today. Using the sum over histories, they can calculate the precise shape of the primordial spectrum, which can later be compared to the WMAP measurements.

This simplified model can be easily extended to the M-theory landscape. Although the landscape accepts a large number of no boundary alternative histories, the solution to the wave equation in which the field rolls down from a wide, flat plateau leads to flat universes like our own. The formation of stars and galaxies requires that the primordial quantum fluctuation be sufficiently large. In the top-down model, only a few potentials meet those stringent requirements. According to the Hawking-Hertog proposal, a few of the possible valleys in the landscape will have significant probabilities.

The Observer-Selected Universe

At this point, the Euclidean quantum gravity cosmology described by the top-down model of Hawking and Hertog is still in development. To obtain the wave function that can describe the present quantum state of the universe, top-down cosmology uses Feynman's sum over histories method with only final boundaries—a method that can be achieved only in Euclidean time. Since the model does not use initial boundaries, in Euclidean top-down cosmology as well as in the earlier no boundary proposal, the universe is self-contained and has no origin.

But what really is Euclidean time? As I explained in chapter 9, Euclidean time is a mathematical tool in which time, expressed in imaginary numbers, behaves just as a spatial dimension, increasing or decreasing in value just like any spatial dimension. In the Euclidean approach, instead of a spacetime of three spatial dimensions and one of time, there are four spatial dimensions. The no boundary proposal is rooted in Euclidean time. The absence of singularities at the origin of the universe works only in Euclidean time. When we return to the real time that we experience in our universe, the singularities return. This problem has prompted Stephen Hawking to suggest that perhaps "the so-called imaginary time is really the real time, and that what we call real time is just a figment of our imaginations. . . . So, maybe what we call imaginary time is really more basic, and what we call real [time] is just an idea that we invent to help us describe what we think the universe is like."[13] For Hawking, the two concepts are equally valid. Since a scientific theory is a mathematical construction that we create to help us understand the world, it only exists in our minds. "So it is meaningless to ask: Which is 'real' and which is 'imaginary' time? It is simply a matter of which is the more useful description."[14]

For Hawking and Hertog, the central problem of cosmology is resolved, since present-day observers, human or not, *select* the history of the universe. Thus, the histories of the universe depend on the observations that we make today. The observed value of the cosmological constant, for example, is chosen as a final boundary for one of the histories. "A top-down approach," says Hertog, "does not tell us *how the universe should be,* but *why the universe is the way it is.*"[15]

Hawking and Hertog's simplified model shows that top-down cosmology predicts universes that rapidly inflate and later expand, to end up with perfectly flat geometries, as in the universe that we observe. As with all inflationary models, primordial quantum fluctuations have been magnified during the rapid inflation to become the seeds for the observed structure of the universe. Neighboring valleys in the M-theory landscape can also have significant probabilities.

The shape of the primordial spectra of energy fluctuations as well as the spectrum of gravitational waves depend on the specific geometry of the top-down model. This dependence can be tested against the WMAP observations and the future Laser Interferometer Space Antenna, LISA, to be launched by NASA and the European Space Agency toward the middle of the second decade of this century (Plate XV). Agreement with experimental observations is of course the only valid test for any model.

When and if the top-down approach passes that test, we would know if our observations do indeed select the history of our universe.

A Self-Organizing Universe

Euclidean quantum gravity leads to a different view of the relation between cause and effect. Since Euclidean time behaves like a space coordinate, it can increase or decrease. Without the standard notion of time that flows only in one direction, there is no underlying structure where events can be placed in any specific order, and the concepts of cause and effect lose their meaning. But without cause and effect, the evolution of the universe would disappear. Since the universe does evolve, Euclidean quantum gravity models such as the top-down approach propose that, just as stars and galaxies emerged from the primordial quantum fluctuations, causality would also emerge out of these fluctuations that, by themselves, had no causal structure.

Because calculations in quantum gravity are so complex, some scientists have been working on a parallel path from the theorists, taking advantage of the increasing power of computers to model the structure of spacetime. Using four-dimensional triangular shapes called *four-simplices* as building blocks, these computer models can spontaneously assemble representations of a four-dimensional curved spacetime by following the rules of quantum theory. The models are particularly useful in representing spacetime curvatures at very small-distance scales, the quantum regime where general relativity breaks down.

In this approach, spacetime is built as a sum over geometries, the quantum mechanical superposition of all the possible shapes that can make up space. In the simulations, each shape is made up of a collection of four-simplices. To obtain the superposition, which is a weighted average of all these possible forms, scientists weigh each shape based on the way all the simplices come together to form the shape.

When these simulations are done for the Euclidean quantum gravity models without causality, the superpositions do not cancel out to form a smooth spacetime, as expected. Instead they reinforce each other and either collapse into a small volume with infinite dimensions or expand into large two-dimensional strands. Neither possibility is the universe that we observe.

In 1998, Jan Ambjørn of Copenhagen University, Jerzy Jurkiewicz of Jagiellonian University in Krakow, and Renate Loll of Utrecht University decided to include causality in the simulations by adding the arrow of time to the interactions of each simplex. In 2004, they showed that, for a simple case, the computer generated a spacetime of four dimensions (actually 4.02 dimensions) and not two-dimensional strands or tiny balls of infinite dimensions.[16] With their *causal dynamical triangulations* method, they were able to derive, for the first time, the actual number of large dimensions of our universe from the superpositions of simple shapes.

Encouraged by this initial result, the researchers decided to see if the simulations would generate a universe that agrees with general relativity at large scales. To do that, they needed to provide as input the actual value of the cosmological constant. With this input, the simulations generated a spacetime with a positive curvature but devoid of matter—what physicists call a de Sitter universe.[17] By adding the arrow of time to the interactions of simple four-dimensional geometrical shapes and the present value of the cosmological constant, the de Sitter geometry emerged. Similar collective or *emergent* behavior is seen in other areas. Large flocks of birds, for example, behave in a collective way without following any leader, as individual birds react only to the behavior of nearby birds.

Ambjørn, Jurkiewicz, and Loll went on to investigate the structure of spacetime by studying the diffusion process. Diffusion occurs when two substances mix, increasing the entropy of the system. By studying the way one substance diffuses into the other, we can gain information about the underlying structure of the substances. Applying the diffusion equations to the superposition of geometries, they could see how the spreading evolved as a result of quantum fluctuations. What they found was remarkable: the number of dimensions of spacetime depends on the scale (Fig. 10–3). At small scales, up to about 10^{-34} meter, spacetime has less than four dimensions. At shorter distances, quantum fluctuations of spacetime dominate and classical geometry breaks down. At that tiny scale, the number of dimensions steadily drops to about two but remains continuous, not foamy as Wheeler had proposed.

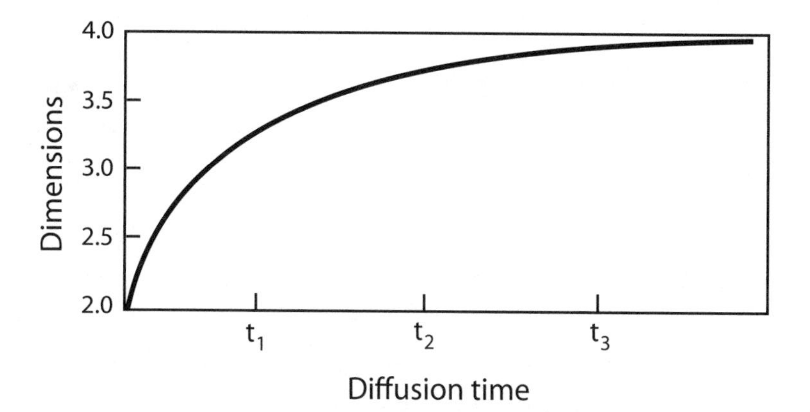

Figure 10-3. At very small scales, below 10^{-34} meter, the number of dimensions of spacetime drops from four to about two, one spatial dimension plus time.

Much remains to be done. The simulations still don't include matter. That should not be an easy task, but until that is done, testable predictions cannot be drawn from the model to validate it as a theory of quantum gravity.

The possibility that the universe self-organized spontaneously will bring us closer to solving the problem of origins. The model, however, still requires the preexistence of the arrow of time and the value of the cosmological constant.

The top-down Euclidean quantum model of Hawking and Hertog gives us a model of a universe without origin in Euclidean time that selects its own history through the observations of intelligent observers. The cosmological constant, the number of dimensions, and the laws of physics, including the arrow of time, are final boundaries selected by observers and become part of the history of the universe. On the other hand, the emergent model of Ambjørn, Jurkiewicz, and Loll gives us a self-organizing universe in real time that does not explain the arrow of time or the cosmological constant. But the initial simulations already attempted show that a Euclidean quantum model does not generate the real universe. Clearly, more work is needed on both models before we can take sides.

Does the problem of origins remain unresolved? Does the universe need a supernatural creator? Let's look further.

Chapter 11

ORDER WITHOUT DESIGN

PREDICTING THE EARLY UNIVERSE

We know with as much certainty as science can give us that our universe—our pocket universe or our cycle in the infinite chain of universes—began 13.7 billion years ago in the big bang. The discovery of the details of the first moments of the universe ranks as one of the greatest scientific discoveries of the twentieth century—a triumph of creativity, ingenuity, imagination, and perseverance, with a certain element of serendipity. With Einstein's general theory of relativity as the main tool, Alexander Friedmann and George Lamaître proposed for the first time in history a scientific model of a dynamic universe. Two years before Edwin Hubble's dramatic experimental observation of the expansion of the universe, Lamaître had concluded that the dynamic universe of Einstein's equations was evidence for the universe's origin.

Using his knowledge of nuclear physics, George Gamow proposed the correct mechanism for the formation of the light elements in the process known as nucleosynthesis, which took place during the first ten minutes after the big bang. In 1956, Gamow and his collaborators Ralph Alpher and Robert Herman calculated that when the process of nucleosynthesis ended, the universe entered a plasma phase that lasted for 300,000 years, at the end of which the universe became transparent to light and the first atoms were formed. Based on these calculations, they predicted the existence of a primordial radiation, the cosmic microwave background radiation, which was later detected by Arno Penzias and Robert Wilson in 1965 and, more recently, was spectacularly confirmed by NASA's COBE satellite. The COBE results told us unequivocally that the universe had indeed started in a big bang event, as Lamaître had envisioned, and that Gamow's predictions about the nucleosynthesis of the light elements and the existence of the primordial radiation were correct. The confirmation of the predictions of the big bang theory prompted Stephen Hawking to proclaim it "the discovery of the century, if not of all time."

But as we have seen in the previous chapters, today we know much more. We know the exact details of the evolution of the early universe from a time so close to the actual beginning that even the shortest intervals of time in modern computer processors, with their clocks running at gigahertz frequencies, seem like eons in comparison. A five-gigahertz processor, for example, has a clock that oscillates 5 billion times in 1 second with each oscillation lasting for 0.2 nanosecond, or 0.2×10^{-9} second. We can trace the events of the early universe up to the end of the inflationary era, 10^{-35} second after the big bang. That's twenty-six orders of magnitude smaller!

Twenty-five years before NASA's WMAP instruments corroborated it, Robert Dicke predicted in a lecture at Cornell University that the universe had to be flat within one part in 10^{14} or it would have either collapsed soon after its birth or it would have expanded before having a chance of forming galaxies. Dicke was troubled by his own prediction, because for the universe to be flat, a very delicate fine-tuning was required and he didn't see a way to explain it. Almost exactly one year later, partly motivated by Dicke's lecture, Alan Guth proposed his successful inflationary theory of the universe, which provided a mechanism to solve the flatness problem and resolve Dicke's fine-tuning paradox.

But a universe that is too flat doesn't work because it can't make stars and galaxies. It has to have some regions that protrude or dip here and there so that gravity can magnify those irregularities and make stars. The astonishing prediction made by the physicists who met at the Nuffield Workshop on the Very Early Universe that Hawking organized the following year is, in my opinion, one of the most striking examples of the predictive power of science. It is very unfortunate that the general public remains at the margin of these events—due of course to their technical nature—and misses out on how a construct of the human mind that is "only a theory" can predict with uncanny accuracy events that happened when no planets, stars, galaxies, or even molecules existed.

As detailed in chapter 4, the physicists gathered at Nuffield used the equations of general relativity, quantum mechanics, and pure mathematics to come up with the fingerprint of the early universe. What is most interesting is that none of the scientists gathered there thought for a moment that their predictions would be corroborated in their lifetimes or perhaps in their children or grandchildren's lifetimes.

At Nuffield, the physicists calculated the spectrum of density perturbations by applying the equations of general relativity to the inflationary model. Using several mathematical methods, they obtained an expression for a scale-invariant spectrum with a tilt. Their calculations showed that the quantum fluctuations during the early universe should appear random at first sight but could actually be decomposed in simple sine waves with a very specific variation in the wavelengths (the tilt). With this exact pattern, the physicists and cosmologists at Nuffield predicted that the early universe could evolve to make stars and galaxies.

Only ten years after the workshop, in 1992, NASA announced that its COBE satellite had measured the spectrum of the early universe and that it matched the calculations made at Nuffield. However, the instruments aboard COBE couldn't measure the tilt. That measurement had to wait for the more advanced instruments on WMAP. In 2006, the WMAP science team announced that the satellite's delicate measurements agreed completely with the predictions of the inflationary theory, tilt and all.

It is nearly impossible to obtain such a striking agreement between the predictions of the theory and the delicate and precise experimental measurements and not have the right theory. The inflationary theory is correct. It is not yet complete, but its basic premise is correct. And so are general relativity, nuclear physics, and the laws of physics, which were all used to develop the detailed predictions of the structure of the CMB.

The Origin of Complexity

Our universe had an origin and we can explain with the luxury of detail how it evolved from an instant after its creation up to the present day.

At the beginning, the universe was simple, and the processes were few. As it evolved, it grew in complexity and our detailed knowledge of all the processes decreased. We know more about the first minute of the universe than about the next billion years. But we know enough to understand the main steps that it took to get here.

If the universe was simple earlier and complex now, how did it develop its complexity? Recent studies on the behavior of complex systems are beginning to provide additional support for the idea that the complexity that we observe can arise automatically. In the previous chapters, we have seen that, given the laws of physics and the values of the fundamental constants, the evolving universe can develop complexity. Although at first sight the laws of physics appear to be very complex, they are actually fairly simple. Simple rules can lead to complex systems. The algebraic rules to generate the Mandelbrot fractal shown in Plate X are very simple, yet the structure of the fractal itself possesses a rich structure. As I explained in chapter 7, the endless generation of pocket universes in the eternal inflation scenario can be described by the self-similarity at all scales exhibited by a fractal.

The study of complex behavior is usually illustrated with the device known as a *cellular automaton*, a model developed in the 1940s by Stanislaw Ulam and John von Neuman while they were at Los Alamos National Laboratory working on the Manhattan Project.[1] Ulam had been interested in crystal growth and von Neuman had worked on the problem of self-replicating systems. Von Neuman's motivation was the problem of one robot building another robot. Ulam suggested that they apply the mathematical ideas that he had developed for crystal growth to the robot replication problem. The result was a universal copier and constructor, the first cellular automaton. After Ulam and von Neuman's pioneering work, research on cellular automata continued at MIT, with Edward Fredkin and his collaborators.

A cellular automaton can be simulated on a grid with a set of rules that apply to the state of each cell in the grid. In the case of a two-dimensional cellular automaton, each cell has eight immediate neighbors. If each cell has two states, 0 and 1, or black and white, the nine neighbor cells would have 2^9 or 512 states that give rise to a large number of patterns. Each one of these patterns has a rule for the state of the center cell. In 1970, John Horton Conway at Princeton invented a cellular automaton called "The Game of Life," which became extremely popular. Its rule, like that of all automatons, is simple: If a black cell has two or three black neighbors, it remains black, otherwise it switches to white; if a white cell has three black neighbors, it switches to black, otherwise, it remains white. After applying the rule, preferably with a computer, the automaton generates a complex behavior. Plate XVI shows a complex pattern that I generated using a similar simple rule.

The study of the generation of complexity from simple systems is still in its infancy. Much work needs to be done before testable models can be constructed. Although a cellular automaton with a simple algebraic rule can generate a rich structure that exhibits complex behavior in relatively few steps, these cases remain only demonstrations of the generation of complexity, not simulations of real systems. But they serve to indicate that, given ample time, the complexity of the present universe can be explained.

The Problem of Origins One More Time

Our universe had an origin and it seems plausible that with its simple laws of physics, it generated the complex world we see out the window. But we are seeking an answer to the ultimate question: Where did our universe come from? However, our universe may not be all that there is. Our universe might be a pocket universe in a vast landscape of other pocket universes that populate a multiverse. Or it might be one of an infinite series of cycles. If this is the case, the problem of the origin of our universe is then transferred over to the problem of the origin of the multiverse. This is the problem of origins.

Intimately related to the problem of origins is the question of whether science has explained the fragile equilibrium in the laws of physics that makes possible our existence. For example, has science explained the precise fine-tuning of the mechanisms in the sun that allow an energetic gamma-ray photon produced in its interior to undertake myriad collisions during one million years before emerging through its surface to rush toward our delicate retinas? Or the even more unlikely fine-tuning of the fine structure constant that controls the existence of an excited state in carbon-12? This excited state matches exactly the combined energies of a beryllium and a helium atom colliding in the interior of a star. As a result, these atoms stick together only 4 out of 10,000 times so that the relatively few carbon atoms that form can decay to their ground states. And once in the ground state, carbon can form the complex organic molecules that make up our bodies. Are these mere coincidences or was the universe designed for life?

The question of the fragile equilibrium between the mechanisms in the sun and our existence isn't difficult to answer. Our universe is vast. Scientists estimate that the number of stars in our Milky Way galaxy alone is of the order of 100 billion. And there are at least as many galaxies in the observable universe. We know that many stars in our galaxy alone are G2 stars, just like our sun, emitting a very similar spectrum of radiation. Many of those stars contain planetary systems of their own, and we are beginning to discover them (Plate XI). It is easy to see that our universe must contain many solar systems with the right conditions for our existence. Our solar system happens to be one of those, and we live here.

However, the question of the value of the fine structure constant cannot be resolved as easily. This constant is a combination of several of the fundamental constants in our universe. The question of why it has the value that it has and no other is actually a question of why the fundamental constants are what they are.

An even more difficult question to answer is that of the small positive value of the cosmological constant or the energy of the vacuum. As explained in chapter 6, the experimental discovery of the acceleration in the expansion of the universe 5 billion years ago shows that this acceleration requires that all vacuum energy contributions cancel out to 120 decimal places.

The problem of origins is then threefold: first, we need to understand how the universe—or the multiverse—came to be; second, we need to understand the origin of the laws of physics; and third, we need to explain the fine-tuning that we observe. We want a satisfactory answer to all three problems. How do the models that we have considered stand on these three issues?

First, a word of caution regarding these models: they are not full-fledged theories since they aren't yet part of a self-consistent theory of quantum gravity. We must, therefore, take them as works in progress, some of them with a great deal of promise.

The eternal inflation scenario tells us that an infinite number of universes will randomly and spontaneously spring out of each one of the 10^{500} possible stable vacuum configurations of the landscape. Eternal inflation is then eternal into the future. Is it also eternal into the past? Does the infinite chain of universes extend infinitely in the past or does it encounter a singularity? Alexander Vilenkin of Tufts University and Alvin Borde looked at this issue in the 1990s and concluded that future models and expansions to the current eternal inflation model could solve the problem of the initial singularity.[2] In 2002, Anthony Aguirre of the Institute for Advanced Studies and Steven Gratton of Princeton University took a step in that direction; they proposed a mechanism that could form the basis for a future eternal inflation model without a singularity at the beginning.[3] Such models have not yet been proposed and, as the authors state, they may not be problem free and may be very difficult to develop.

In the current eternal inflation model,[4] the fundamental laws of physics must be in place at the origin and throughout all the phases of the multiverse, although individual pocket universes will have different values of the constants of nature. If Aguirre's mechanism is successfully implemented, the model will describe an eternal universe into the past as well and will solve the problem of the origin of the universe. However, even if the mechanism is successful, the current model does not resolve the issue of the fine-tuning of the cosmological constant without resorting to the anthropic principle. Several groups are currently attempting to develop extensions to the model that could explain the small value of the cosmological constant.[5]

The cosmological natural selection model offers an alternative and perhaps complementary mechanism for the generation of pocket universes in the landscape.[6] However, the model does not address the issue of the creation of the multiverse. A more serious problem with this model is that since, according to quantum mechanics, the total information of the universe must remain constant, information cannot leak out of the universe through a wormhole to another universe, a process that is at the heart of the model.

The cyclic model offers the possibility of an eternal universe that has been going through trillion-year cycles of expansion and contraction with no beginning and no end. Each cycle contains its own set of the laws of physics and constants of nature. During each cycle, the laws of physics can take any possible form and the constants of nature as well as the cosmological constant can have any value. However, the action of the proposed Abbott quantum field acting over the infinite series of past cycles automatically and mindlessly drives the laws of physics toward the form that we observe today and the constants of nature toward the values that we measure.[7] More important, Abbott's field produces the small and positive cosmological constant that we observe in our present universe.

With the laws of physics and the constants of nature in place, each cycle gives birth to a new universe that evolves stars, galaxies, planets, and life. The quantum mechanical fluctuations that give rise to the structure in each cycle take place at the end of the previous cycle. However, it has been argued that it is precisely those quantum mechanical fluctuations that may prevent an infinite number of previous cycles, since quantum fluctuations are random and unpredictable. There is some probability that, due to their random nature, they might cause an interruption of the cycles.

With the work of Paolo Creminelli of the Abdus Salam International Center in Italy and Leonardo Senatore of Harvard University to replace the collision of the oscillating branes in the cyclic model for a smooth transition that forces the branes to move apart just before they collide, the scale-invariant spectrum generated by quantum fluctuations can pass unscathed to the new cycle.[8] Although in their initial attempt, the scale-invariant spectrum they obtained had the wrong tilt; a second group at the Perimeter Institute for Theoretical Physics in Canada used similar techniques to obtain a spectrum with the correct tilt.[9] These two scenarios offer some promise and may restore the model's claim of an eternally oscillating universe.

The self-organizing model described in the previous chapter is an exciting exercise in the computer modeling of spacetime and indicates the possibilities of this kind of work. The model shows that spacetime can spontaneously organize itself as a sum over geometries using shapes formed by four-dimensional triangles. Imputing the arrow of time and the current value of the cosmological constant into the requirements for the calculations, the shapes generate a de Sitter universe with the correct number of dimensions. However, the model requires the existence of the laws of physics, the arrow of time, as well as the cosmological constant, and, at this stage, is not yet a cosmological proposal or model.

Finally, the Euclidean quantum gravity top-down approach presents a model of an observer-selected universe without an origin in Euclidean time based on the no boundary proposal developed earlier by Hartle and Hawking. The model explains the fine-tuning of the cosmological constant and every other observed constant of nature because these parameters are final boundaries that are selected for the model. Since the laws of physics give rise to the parameters that are observed today, they are indirectly selected by present-day observers. Although the universe had no origin in Euclidean or imaginary time, it does in real time. As described in chapter 10, Hawking has argued that, since any scientific theory is a mathematical construction, the two concepts—real time and Euclidean time—are equally valid and the question is not about which one is real, but about which one is more useful.

A Universe without a Designer

We know that our present universe had an origin in the big bang. We also know that it is possible for science to explain how this universe came to be, although we don't have the definite answer yet. Whether our universe is a pocket universe amid an unimaginably large number of other pocket universes that have been coming into existence and evolving for an eternity, a cycle in an infinite series of cycles, or an observer-selected universe without an origin in real time, we cannot yet know. The three models behind these possibilities give us a universe without an origin, a universe that does not need to be set in motion. As Stephen Hawking speculated, referring to his earlier no boundary proposal, in a universe with no origin there would not be a place for a creator.[10]

Do the models solve the problem of origins in full? Or does the creator still have a job to perform? Before addressing these questions, I think that it is important to emphasize once more that the models, proposals, scenarios, and constructs that we have examined are just what these names imply: tentative but serious and precise models that have not yet reached the status of a theory, but that are being carefully developed and discussed in an effort to understand the world. Eternal inflation is a relatively mature model, but the Aguirre mechanism that describes a universe without origin has not been fully incorporated within it. Moreover, eternal inflation has yet to be integrated with M-theory. Whether this is a problem or an advantage, we won't know for some time since no one yet knows if M-theory is the correct theory of matter.[11] The newer cyclic model is gaining maturity, but the issue of the smooth transition of the quantum fluctuations between cycles is still in development. And since the cyclic model is based on M-theory, it will have to wait for the verdict on that theory. Finally, the Euclidean quantum gravity approach is still in its very early stages of development. And before any of these models evolve into complete theories that can be corroborated by observation with the same confidence as that of the big bang theory, a complete theory of quantum gravity must be developed.

On the other hand, it is important to emphasize that science has an impressive track record, as shown by the astonishing agreement of the predictions of the big bang and inflationary theories of the evolution of the early universe with the precise and delicate data collected by COBE and WMAP. The extraordinary success of these theories—which are correct but still incomplete—shows that scientists have been able to fully resolve many instances of fine-tuning that initially looked daunting, such as the flatness and horizon problems.

Where do the models that we are considering here stand regarding the full solution of the problem of origins, then? As stated earlier, the three models with their recent extensions give us a universe without origin. However, in eternal inflation, the laws of physics, the values of the fundamental constants, and the small value of the cosmological constant can be explained only by resorting to the anthropic principle. Unless recent work on extensions to the model can provide a viable mechanism for these issues, eternal inflation, in its present form, still gives us an unsatisfactory solution to the problem of origins. The Euclidean quantum gravity top-down approach does solve the problem of origins fully since the history of the universe is selected by the measurements of present-day observers. The cyclic model also solves the problem of origins with a series of cycles that is infinite into the past and a proposal that explains the laws of physics, the values of the constants of nature, and the fine-tuning of the cosmological constant.

Is cosmology on the right track with these models? We should get an answer to this question in the not-too-distant future. All three models make predictions that are testable. With the new particle accelerators coming on line, such as the Large Hadron Collider at CERN already in operation, the planned orbital observatories, as well as more accurate techniques, those tests can be conducted. Observation and measurement are the arbiters that will determine the outcome, and we must wait for their verdict.

Clearly, there are obstacles that the models must overcome: scientists still need to develop them into full-fledged theories and they must pass the test of experimental observation. However, there is little doubt that science can explain the universe, as evidenced by the extraordinary advances in our understanding of the evolution of the early universe right up to an instant after the big bang. If one of the present models or a more advanced one yet to be developed turns out to be the correct one, the problem of origins would be fully explained and the creator wouldn't have a job to do. The universe and its laws of physics would have no origin and would not need a supernatural designer. The fine-tuning observed would be the result of the laws of physics—the universe's watchmaker—that evolved purposelessly and mindlessly to create the equilibrium and order that we see today.

NOTES

1. DESIGNER UNIVERSE

1. William Paley, *Natural Theology or Evidences for the Existence and Attributes of the Deity Collected from the Appearance of Nature* (London: Halliwell, 1802).

2. Ibid., p. 2.

3. A kelvin (symbol K) is the unit of temperature in the absolute scale, which starts at zero, the temperature at which no thermal energy can be extracted from an object. Absolute zero corresponds to –273.15 °C or –459.67 °F.

4. Richard Dawkins, *The Blind Watchmaker* (New York: W. W. Norton, 1996).

5. Albert Einstein, "Notiz zu der Arbeit von A. Friedmann 'Über die Krümmung des Raumes,'" *Zeitschrift für Physik* 16 (1923): 228.

6. He died of some serious illness, "probably typhoid fever, and died in a state of delirium." Simon Singh, *Big Bang* (New York: Fourth State, 2004), p. 155.

7. Augustine, *City of God*, in W. T. Oates, ed., *Basic Writings of Saint Augustine* (New York: Random House, 1948), p. 6.

8. Hawking and Hartle's description employs *Euclidean time*, a mathematical tool used in quantum mechanics in which the time coordinate is expressed as one of the spatial dimensions. Hawking proposes that real time and Euclidean time are equivalent and that whichever one we use is a matter of convenience. See Stephen Hawking, *A Brief History of Time* (New York: Bantam Books, 1988), p. 139.

9. Ibid., p 136.

10. S. W. Hawking and T. Hertog, "Populating the Landscape: A Top Down Approach," *Physical Review* D73 (2006): 123527.

11. "If the universe is really completely self-contained, having no boundary or edge, it would have neither beginning nor end: it would simply be. What place, then, for a creator?" Hawking, *Brief History of Time*, p. 141.

2. BUILDING A THEORY OF THE UNIVERSE

1. Quoted in D. Sobel, *Galileo's Daughter* (New York: Walker Publishing Company, 1999), p. 244.

2. In the Gregorian calendar, in use in Italy and elsewhere on the Continent since 1582, Newton's birthdate falls ten days later, on January 4, 1643. However, the Gregorian calendar was not in use in England at the time.

3. University Library, Cambridge, Additional Manuscript, 4007, ff., 707^{r}-7-7^{v}. Quoted in G. E. Christianson, *In the Presence of the Creator: Isaac Newton and His Times* (New York: Free Press, 1984), p. 283.

4. Isaac Newton, *Unpublished Scientific Papers of Isaac Newton: A Selection from the Portsmouth Collection in the University Library, Cambridge* (Cambridge: Cambridge University Press, 1962), p. 121.

5. Isaac Newton, *Principia Mathematica* (Berkeley: University of California Press, 1934), p. 13.

6. Ibid., p. 550.

7. Ibid., p. 13.

8. Ibid. Newton introduced three laws of motion. The third law—which I didn't mention here because it isn't directly relevant for the discussion—says: To every action there is always an equal and opposite reaction: or, the mutual actions of two bodies upon each other are always equal and directed to contrary parts.

9. Albert Einstein, *Autobiographical Notes*, in A. Schilpp, *Albert Einstein: Philosopher-Scientist* (LaSalle, IL: Open Court, 1982), p. 19.

10. Albert Einstein, "Cosmological considerations on the General Theory of Relativity," English translation in Albert Einstein, *The Principle of Relativity* (New York: Dover, 1952), pp. 177–88.

11. Ibid., p. 183.

12. Although Newton himself believed that the universe was finite, the physics constructed from his work, what we call today *Newtonian physics*, allows an infinite universe.

13. Einstein, "Cosmological Considerations," p. 185.

14. Galileo Galilei, *Dialogues concerning the Two Chief World Systems* (Berkeley: University of California Press, 1967), p. 186.

15. From Einstein's 1922 lecture in Kyoto, reproduced in T. Ogawa, *Japanese Studies in the History of Science* 18 (1979): 73.

16. Ibid., p. 78.

17. Albert Einstein, "Über das Relativitätsprinzip und die aus demselben gezogenen Folgerungen [On the Principle of Relativity and the Conclusions

Drawn from It]" *Jahrbuch der Radioaktivität und Elektronik* [*Annual Review of Radioactivity and Electronics*] 4 (1907): 411–62.

18. A. S. Eddington, *Space, Time, and Gravitation* (New York: Cambridge University Press, 1920), pp. 114–16.

19. C. W. Misner, K. S. Thorne, and J. A. Wheeler, *Gravitation* (New York: W. H. Freeman, 1973), p. 5.

20. Albert Einstein, *Relativity: The Special and General Theory* (New York: Crown Publishers, 1961), p. 108.

21. Einstein, "Cosmological Considerations," p. 180.

3. In the Beginning

1. Einstein to George Gamow, as described in "The Evolutionary Universe," *Scientific American*, September 1956. Reproduced in *Cosmology + 1* (San Francisco: W. H. Freeman, 1977), p. 16.

2. Albert Einstein, "Cosmological Considerations on the General Theory of relativity," English translation in *The Principle of Relativity* (New York: Dover, 1952), p. 180.

3. Albert Einstein, *The Meaning of Relativity* (Princeton, NJ: Princeton University Press, 1922), p. 112.

4. A. A. Friedmann, *Mir kak prostanstvo i vremya* [The World as Space and Time] (Moscow: Nauka, 1965). French translation in *Essais de Cosmologie, textes choisie, traduits due russe el l'anglais et annotes* (Paris: Seul, 1997), p. 38.

5. G. Lamaître, *The Primeval Atom* (New York: Van Nostrand, 1950), p. 77.

6. The Doppler shift is the same technology used by law enforcement officers to measure car speeds and by weather forecasters to measure the speed of precipitation.

7. W. D. Ault, "Oxford in 1907 (With a Glimpse of T. E. Lawrence)," *American Oxonian* 66, no. 2 (1979): 122.

8. G. E. Christianson, *Edwin Hubble: Mariner of the Nebulae* (Chicago: University of Chicago Press, 1995), p. 64.

9. Ibid., p. 65.

10. M. L. Humason, "Edwin Hubble," *Monthly Notices of the Royal Astronomical Society* 114, no. 3 (1954): 291.

11. Edwin Hubble, "A Relation between Distance and Radial Velocity among Extra-Galactic Nebulae," *Proceedings of the National Academy of Science* 15 (1929): 168–73.

12. Hubble's law is velocity = $H \times$ distance, where H is the Hubble constant. Simple algebra gives H = velocity/distance. Velocity is defined as distance/time (we say, for example, that we're traveling at 90 kilometers per hour). We can write for the Hubble constant: H = (distance/time)/distance = distance/(time × distance). Distance cancels out and we're left with $H = 1/\text{time}$ or time = $1/H$ (called *Hubble time*). Thus, the time of the expansion of the universe (the age of the universe) is given by the inverse of Hubble constant.

13. In addition to the proton and the electron, the beta decay process generates another particle called the *neutrino*. Neutrinos have no electric charge, travel at speeds close to the speed of light, and have a tiny mass that is too small to be measured accurately with current methods.

14. R. A. Alpher, H. Bethe, and G. Gamow, "The Origin of Chemical Elements," *Physical Review* 73 (1948): 803.

15. G. Gamow, "Expanding Universe and the Origin of the Elements," *Physical Review* 74 (1948): 505.

16. The microwave region of the electromagnetic spectrum has wavelengths ranging from about 1 mm to 1 m. A microwave oven operates at a wavelength of 12.2 cm. The microwave region overlaps the ultra-high frequency (UHF) band, which ranges from 10 cm to 1 m. Wireless phones operate in several radio regions. One earlier type, still in use, operates in the 2.4 GHz frequency, which corresponds to a wavelength of about 12 cm, that of microwave ovens, and interferes with their signals when they are in operation. A newer model at 5.8 GHz or about 5 cm in wavelength has fewer interference issues.

17. R. A. Alpher, J. W. Follin Jr., and R. C. Herman, "Physical Conditions in the Initial States of the Expanding Universe," *Physical Review* 55 (1939): 434.

18. Although it was Lamaître who came up with the idea for the big bang and Gamow who established it as a scientific theory, the name "big bang" wasn't invented by any of them. It was the British astronomer and cosmologist Fred Hoyle, author of a rival cosmological theory of continuous creation, who coined the term during a radio broadcast in the 1940s to ridicule the theory. Hoyle's own continuous creation theory was shown to be incorrect some years later.

19. A. A. Penzias and R. W. Wilson, "A Measurement of Excess Antenna Temperature at 4080 Mc/s," *Astrophysical Journal* 142 (1965): 419–21; R. H. Dicke, P. J. E. Peebles, P. G. Roll, and D. T. Wilkinson, "Cosmic Black-Body Radiation," *Astrophysical Journal* 14 (1965): 414–19.

20. Spacecraft grounding is a pesky issue. After more than 120 missions, we're still finding and correcting grounding problems in the space shuttle.

21. Today, John Mather is the chief scientist for the Science Mission Directorate at NASA Headquarters in Washington, DC.

22. G. Gamow, "The Origin of the Elements and the Separation of Galaxies," *Physical Review* 74 (1948): 572–73; Ralph A. Alpher and Robert Herman, "Evolution of the Universe," *Nature* 162 (1948): 774–75.

4. How to Make a Universe

1. Alan Guth, *The Inflationary Universe* (Reading, MA: Addison-Wesley, 1997), p. 21.

2. That year's selection for the physics prize was somewhat discussed, since no experimentalist had won the Nobel Physics Prize until then. Moreover, Penzias and Wilson did not design and set up an experiment to make their discovery; they simply stumbled upon it. The other reason why the 1979 prize was somewhat controversial was that neither Dicke nor Peebles were included to share the prize, even though their contribution to the discovery was extremely important.

3. The esoteric particles are the magnetic monopoles. These particles have never been observed, in spite of the heroic efforts of several researchers a few years ago. A regular magnet has two poles, a north and a south. These magnetic poles exist only in pairs and if you attempt to separate them by splitting a magnet in two, you end up with two new full magnets, each with its own north and south poles. The process can continue all the way to atomic dimensions without ever isolating a single pole. In 1931, the English physicist Paul Dirac postulated that single magnetic poles, or monopoles, should exist to round off the symmetry between electricity and magnetism. In 1974 Gerald 't Hooft in the Netherlands and Alexander Polykov in Russia independently suggested that certain physics theories imply the existence of these particles. Magnetic monopoles are thought to be extremely heavy, about 10^{17} times as heavy as protons.

4. In reality, the NASA logo and the shuttle's name break the symmetry.

5. P. G. Debenedetti and H. E. Stanley, "Supercooled and Glassy Water," *Physics Today* 56 (June 2003): 40–46.

6. Guth, *Inflationary Universe*, p. 172.

7. The mass of air in an average-size room, for example, is very small and contributes very little to gravity. The pressure of the air contributes less than a hundred billionth of the small amount that the mass contributes.

8. Guth, *Inflationary Universe*, p. 179.

9. Alan Guth, "The Inflationary Universe: A Possible Solution to the Horizon and Flatness Problems," *Physical Review* D23 (1981): 347–56.

10. A. Lightman and R. Brower, *Origins: The Lives and Words of Modern Cosmologists* (Cambridge, MA: Harvard University Press, 1990), p. 485.

11. A. D. Linde, "A New Inflationary Universe Scenario: A Possible Solution of the Horizon, Flatness, Homogeneity, Isotropy, and Primordial Monopole Problems," *Physics Letters* 108B (1982): 389–92.

12. A. Albrecht and P. J. Steinhardt, "Cosmology for Grand Unified Theories with Radiatively Induced Symmetry Breaking," *Physical Review Letters* 48 (1982): 1220–23.

13. Simon Singh, *Big Bang: The Origin of the Universe* (New York: Fourth Estate, 2004), p. 457.

14. In science, no theory is ever proven to be correct. However, when several important predictions of a theory are validated, the theory gains strength. Most scientists refer to a strong theory as being "correct," but, perhaps not complete. Newton's theory is correct but incomplete.

15. *Time*, February 24, 2003. John Bahcall passed away in his sleep from a rare blood disorder on August 17, 2005. Sadly, he did not live to witness the second and even more exciting WMAP announcement in 2006.

5. THE WATCHMAKER

1. W. Heisenberg, "On the perpetual content of quantum theoretical kinematics and mechanics," *Zeitschrift für Physik* 43 (1927): 172–98, English translation in J. A. Wheeler and W. Zurech, *Quantum theory and measurement* (Princeton, NJ: Princeton University Press, 1983), pp. 62–84.

2. W. Heisenberg, "Ueber die Gundprizipien der 'Quantenmechanik,'" *Forschungen und Forschritte* 3 (1927): 83. Quoted in David Cassidy, *Uncertainty: The Life and Science of Werner Heisenberg* (New York: Freeman, 1992), p. 226.

3. P. A. M. Dirac, interview in *Archives for the History of Quantum Physics* (New York: American Institute of Physics, 1963), p. 30.

4. Electrons are negatively charged. The charge has been measured to be -1.652×10^{-19} coulombs, with the coulomb being the standard unit of charge. The magnitude of this charge, that is, the number without the sign, is the fundamental unit of charge.

5. S. L. Glashow, *The Charm of Physics* (New York: American Institute of Physics, 1991), p. 137.

6. Physicists use the Greek letter nu (ν) to represent neutrinos. The subindexes e, μ (mu), and μ (tau) are used to differentiate the three different kinds of neutrinos.

7. P. W. Higgs, "Broken Symmetries, Massless Particles and Gauge Fields," *Physics Letters* 12 (1964): 132.

8. F. Englert and R. Brout, "Broken Symmetry and the Mass of Gauge Vector Mesons," *Physical Review Letters* 13 (1964): 321; P. W. Higgs, "Broken Symmetries and the Masses of Gauge Bosons," *Physical Review Letters* 13 (1964): 508.

9. H. Georgi and S. L. Glashow, "Unity of All Elementary Particle Forces," *Physical Review Letters* 32 (1974): 438–41.

10. One thousand cubic meters of water contain about 10^{33} protons.

6. Is God in the Details?

1. G. Veneziano, "Construction of a Crossing-Symmetric, Regge-Behaved Amplitude for Linearly Rising Trajectories," *Nuovo Cimento* A57 (1968): 190–97.

2. Y. Nambu, *Proceedings of the International Conference on Symmetries and Quark Models* (New York: Gordon and Breach Publishers, 1970), p. 269; L. Susskind, "Dual Symmetric Theory of Hadrons," *Nuovo Cimento* A69 (1970): 457–96; and K. Zoba and H. B. Nielsen, "Generalized Veneziano Model for *n* Particles, Trajectories with Signature," *Nuclear Physics* B17 (1970): 206–20.

3. J. Sherk and J. H. Schwartz, "Dual Models for Nonhadrons," *Nuclear Physics* B81 (1974): 118–44.

4. String theory officially became superstring theory with the proof of the requirement that supersymmetry was a symmetry of the full ten-dimensional space that appeared in the paper by John Schwartz and Michael Green: M. B. Green and J. H. Schwartz, "Supersymmetrical String Theories," *Physics Letters* B109 (1982): 444–48.

5. M. B. Green and J. H. Schwartz, "Superstring Field Theory," *Nuclear Physics* B243 (1984): 475–536; "Anomaly Cancellation in Supersymmetric D = 10 Gauge Theory and Superstring Theory," *Physics Letters* B149 (1984): 117–22.

6. Although these strings are technically *superstrings*, in most instances, we'll refer to them as strings from here on. Similarly, the current theory is technically called *superstring* theory, but at times it is shortened to *string* theory.

7. T. D. Lee and C. N. Yang, "Question of Parity Conservation in Weak Interactions," *Physical Review* 104 (1956): 254–58.

8. The mirror symmetry is known in physics as *parity conservation*. Parity is a property of the equations that describes the laws of physics and it can be either even or odd. If mirror symmetry holds, parity must not change in the

mirror world: if an interaction starts out with even parity, it must end with even parity, and if it starts with odd parity, it must end with odd parity. The weak interaction violates this rule and as a consequence, parity is not conserved.

9. Physicists at Fermilab have recently reported the results of their analysis of a series of delicate experiments conducted in 2007 in which they found that an unexpectedly large number of muon neutrinos had transformed into electron neutrinos. The scientists cannot come up with any explanation that would explain the phenomenon. An unconventional explanation that had been proposed in 2005 by Heinrich Päs of the University of Dortmund in Germany and his collaborators is now the focus of some attention: a fourth type of neutrino may be oscillating in and out of extra dimensions. A new experiment to verify this idea is now being designed at Fermilab. The new detector is scheduled to be in operation in 2011. If the new experiments show that extra spatial dimensions do exist, it would provide support to M-theory, which postulates the existence of ten spatial dimenions. See Mark Alpert, "A New Kind of Neutrino Hunt," *Scientific American*, September 2008, p. 32.

10. Edward Witten, "Reflections on the Fate of Spacetime," *Physics Today*, April 1996, p. 24.

11. A. S. Eddington, *The Internal Constitution of the Stars* (Cambridge: Cambridge University Press, 1926), p. 292.

12. At those extremely high temperatures, it is the custom to omit the units, since there is little difference among them; 100 million degrees Celsius is about the same as 100 million degrees Fahrenheit or 100 million kelvin.

13. F. Hoyle et al, "A State of C^{12} Predicted from Astrophysical Evidence," *Physical Review* 92 (1953): 1095.

14. S. Weinberg, *Dreams of a Final Theory* (New York: Pantheon, 1992), p. 224.

15. L. Susskind, *The Cosmic Landscape* (New York: Little, Brown, 2006), p. 78.

7. A Landscape of Pocket Universes

1. Scientists have also proposed liquid ammonia as a solvent for cold planets and moons, like some of the satellites of the outer giant planets. There are molecules that can produce compounds at these temperatures, but perhaps not with the efficiency of carbon.

2. L. Susskind, "The Anthropic Landscape of String Theory," arXiv:hep-th/0302219 (2003).

3. R. Bousso and J. Polchinski, "Quantization of four-form fluxes and dynamical neutralization of the cosmological constant," *Journal of High Energy Physics* 6 (2000): 6.

4. Strictly speaking, the term *landscape* wasn't introduced until 2003, in Susskind's paper referenced in these notes. However, Bousso and Polchinski had been working on these ideas for some time.

5. Andre Linde, "The Eternally Existing, Self-Reproducing Inflationary Universe," in *Proceedings of the Nobel Symposium on Unification of Fundamental Interactions*, ed. L. Brink et al. (Singapore: World Scientific, 1987); also published in *Physica Scripta* T15 (1987): 169.

6. A. Borde and A. Vilenkin, "Eternal Inflation and the Initial Singularity," *Physical Review Letters* 72 (1994): 3305–9.

7. A. Borde and A. Vilenkin, "Violation of the Weak Energy Condition in Inflating Spacetimes," *Physical Review* D56 (1997): 717–23.

8. Note that 10^{380} is much, much smaller than 10^{500}. Thus, $10^{500}-10^{380}$ is about equal to 10^{500}. The reader who isn't too familiar with powers of ten may want to try to subtract more manageable numbers that have very different powers of ten, like $10^9 - 10^2 = 1{,}000{,}000{,}000 - 100 = 999{,}999{,}900$. If you had $999,999,900 under the mattress, you wouldn't be lying too much if you say that you're a billionaire (if you're a stickler, you can add the value of the mattress).

9. An American football field is a rectangle 360 feet in length by 160 feet in width.

10. S. Weinberg, "Anthropic Bound on the Cosmological Constant," *Physical Review Letters* 59 (1987): 2607–10.

11. S. Weinberg, *Dreams of a Final Theory* (New York: Pantheon, 1992), p. 229.

12. L. Smolin, "Scientific Alternatives to the Anthropic Principle," arXiv:hep-th/0407213v3 (2004).

13. J. A. Wheeler, "Beyond the End of Time," in *Black Holes, Gravitational Waves, and Cosmology*, ed. Martin Rees, Remo Ruffini, and John Archibald Wheeler (New York: Gordon and Breach, 1974).

14. A. Lawrence and E. Martinec, "String Theory in Curved Spacetime and the Resolution of Spacelike Singularities," *Classical and Quantum Gravity* 13 (1996): 63.

15. M. Bojowald, "Isotropic Loop Quantum Cosmology," *Classical and Quantum Gravity*. 19 (2002): 2717–42; "Dynamical Initial Conditions in Quantum Cosmology," *Physical Review Letters* 87 (2001): 87. I discuss loop quantum gravity in chapter 9.

16. L. Smolin, "Did the Universe Evolve?" *Classical and Quantum Gravity* 9 (1992): 173–91.

8. ETERNALLY OSCILLATING UNIVERSES

1. To get an idea of the length of time for the Planck era, 10^{-43} second, imagine taking a 1-second interval and dividing into a trillion tiny intervals. That's a thousandth of a nanosecond. Take one of these intervals and divide it into a trillion new intervals. Now, take one of these and divide it again into a trillion intervals. And you're not done yet. Take one of these and divide it in a million intervals and, finally, take one of those and divide it into ten intervals. That's the Planck time, 10^{-43} second.

2. P. J. Steinhardt and N. Turok, "A Cyclic Model of the Universe," *Science* 296 (2002): 1436–39.

3. P. Hořava and E. Witten, "Heterotic and Type-I String Dynamics from Eleven Dimensions," *Nuclear Physics* B460 (1996): 506.

4. S. H. Tye and Z. Kakushadze, "Brane World," *Nuclear Physics* B548 (1999): 180–204.

5. E. E. Flanagan, S. H. Tye, and I. Wasserman, "A Cosmology of the Brane World," *Physical Review* D63 (2000): 1103–41.

6. L. Randall and R. Sundrum, "A Large Mass Hierarchy from a Small Extra Dimension," *Physical Review Letters* 83 (1999): 3370–73.

7. Randall and Sundrum use the term *gravitybrane* for what I am calling the Planck brane. The reason for their use of the term will be made clear later.

8. R. C. Tolman, *Relativity, Thermodynamics, and Cosmology* (Oxford: Clarendon Press, 1934).

9. P. J. Steinhardt and N. Turok, *Endless Universe* (New York: Doubleday, 2007), p. 251.

10. L. Abbott, "The Mystery of the Cosmological Constant," *Scientific American* 3, no. 1 (1991): 78.

11. Brian Greene, *The Fabric of the Cosmos* (New York: Alfred A. Knopf, 2004), p. 531.

12. P. Creminelli and L. Senatore, "A Smooth Bouncing Cosmology with Scale Invariant Spectrum," arXiv:hep-th/0702165v1 (2007).

13. N. Arkani-Hamed et al., "Ghost Condensation and a Consistent Infrared Modification of Gravity," *Journal of High Energy Physics* 405 (2004): 74.

14. E. I. Buchbinder, J. Khoury, and B. A. Ovrut, "New Ekpyrotic Cosmology," arXiv:hep-th/0702154v4 (2007).

9. A Universe without Origin

1. J. Wheeler, "On the Nature of Quantum Geometrodynamics," *Annals of Physics* 2 (1957): 604–14.

2. A. Sen, "Gravity as a Spin System," *Physics Letters* B119 (1982): 89–91.

3. A. Ashtekar, "New Variables for Classical and Quantum Gravity," *Physical Review Letters* 57 (1986): 2244–47; and "New Hamiltonian Formulation of General Relativity," *Physical Review* D36 (1987): 1587–1602.

4. T. Jacobson and L. Smolin, "Nonperturbative Quantum Geometries," *Nuclear Physics* B299 (1988): 295–345.

5. C. Rovelli and L. Smolin, "Spin Networks and Quantum Gravity," *Physical Review* D52 (1995): 5743–59.

6. Lee Smolin, *Three Roads to Quantum Gravity* (New York: Basic Books, 2001), p. 186.

7. M. Bojowald, "Dynamical Initial Conditions in Quantum Cosmology," *Physical Review Letters* 87 (2001) 121301-305.

8. S. Hofmann and O. Winkler, "The Spectrum of Fluctuations in Singularity-Free Inflationary Quantum Cosmology," arXiv:astro-ph/0411124v2 (2005).

9. Y. A. Zel'dovich, "The Generation of Waves by a Rotating Body," *Pis'ma v Radakstiyu Zhurnal Eksperimntalnoi i Teoreticheskoi Fiziki* 14 (1971): 270; English version: *Soviet Physics-JETP* 34 (1972): 1159.

10. Stephen Hawking, *Black Holes and Baby Universes* (New York: Bantam Books, 1993), p. 121.

11. Stephen Hawking, "Information Loss in Black Holes," *Physical Review* D72 (2005): 84013.

12. Ibid.

13. D. Christodoulou, "Reversible Transformations of a Charged Black Hole," *Physical Review Letters* 25 (1971): 3552–55; S. W. Hawking, "Gravitational Radiation from Colliding Black Holes," *Physical Review Letters* 26 (1971): 1344–46.

14. J. Beckenstein, "Black Holes and Entropy," *Physical Review* D7 (1973): 2333–46.

15. S. W. Hawking, "Black Hole Explosions?" *Nature* 248 (1974): 30.

16. A. Vilenkin, "Creation of Universes from Nothing," *Physics Letters* 117B (1982): 25.

17. S. Coleman, "Fate of the False Vacuum: Semiclassical Theory," *Physical Review* D15 (1977): 2929–36.

18. J. B. Hartle and S. W. Hawking, "Wave Function of the Universe," *Physical Review* D28 (1983): 2960–75.

19. Stephen Hawking, *A Brief History of Time* (New York: Bantam Books, 1988), p. 136.

20. The first formal publication of Everett's interpretation was a paper in *Reviews of Modern Physics*, which is essentially the same as his doctoral dissertation. The paper citation is: H. Everett, "Relative State Formulation of Quantum Mechanics," *Reviews of Modern Physics* 29 (1957): 454–62.

21. H. Everett, "On the Foundations of Quantum Mechanics," PhD dissertation, Princeton University, 1957.

22. In a book review, Bryce DeWitt writes that Wheeler sat down with Everett and told him what to omit from his dissertation to avoid any problems with the committee. See http://naturalscience.com/ns/books/book02.html. The committee members noted that the dissertation "may be a significant contribution to our understanding of the foundations of quantum theory."

23. Everett's dissertation was thirty-five pages in length. In a footnote, he wrote that he hoped the revised wording would avoid misunderstanding or ambiguity.

24. B. S. DeWitt and N. Graham, eds., *The Many-Worlds Interpretation of Quantum Mechanics* (Princeton, NJ: Princeton University Press, 1973).

25. Max Tegmark, "Parallel Universes," *Scientific American*, May 2003, pp. 41–51.

26. How do we know if a quantum system is coherent or decoherent? By making a measurement on the system. As Seth Lloyd of the Massachusetts Institute of Technology explains, "If a measurement on a quantum system changes its future behavior, then the histories . . . are coherent." See Seth Lloyd, *Programming the Universe* (New York: Alfred A. Knopf, 2006), p. 125.

10. The Self-Selecting Universe

1. D. Page, "Quantum Cosmology," in G. W. Gibbons, E. P. S. Shellard, and S. J. Rankin, eds., *The Future of Theoretical Physics and Cosmology* (Cambridge: Cambridge University Press, 2003), pp. 621–48.

2. Clearly, this sentence wouldn't be "plain English" at a garden party. But it would be for the reader who has reached this far in the book.

3. S. W. Hawking and T. Hertog, "Populating the Landscape: A Top Down Approach," *Physical Review* D73 (2006): 123527.

4. J. A. Wheeler, "The 'Past' and the 'Delayed-Choice' Double-Slit Experiment," in A. R. Marlow, ed., *Mathematical Foundations of Quantum Theory* (New York: Academic Press, 1978), p. 13.

5. The first such experiment was performed by Caroll Alley and collaborators at the University of Maryland. The last was performed by Vincent Jacques and his collaborators in France. The full list of references in chronological order is: C. O. Alley, O. G. Jakubowicz, and W. C. Wickes, "Results of the Delayed Random Choice Quantum Mechanics Experiment with Light," in H. Narani, ed., *Proceedings of the Second International Symposium on the Foundations of Quantum Mechanics*, Tokyo, 1986; T. Hellmuth, H. Walther, A. G. Zajonc, and W. Schleich, "Delayed-Choice Experiments in Quantum Interference," *Physical Review* A72 (1987): 2533; J. Baldzuhn, E. Mohler, and W. Martienssen, "A Wave-Particle Delayed-Choice Experiment with a Single-Photon State," *Zeitschrift für Physik* B77 (1989): 347; B. J. Daku et al., "Delayed choices in atom Stern-Gerlach interferometry," *Physical Review* A54 (1996): 5042; T. Kawai et al., "Realization of a Delayed Choice Experiment Using a Multilayer Cold Neutron Pulser," *Nuclear Instruction Methods* A410 (1998): 259; Y. H. Kim et al., "Delayed 'Choice' Quantum Eraser," *Physical Review Letters* 84 (2000): 1; V. E. Jacques et al., "Experimental Realization of Wheeler's Delayed-Choice GedankenExperiment," *Science* 315 (2007): 966–68.

6. See last reference in the previous note.

7. This arrangement is named after physicists Ludwig Mach and Ludwig Zehnder who invented the design in 1892 to measure the phase shift generated by thin transparent samples placed in the path of one of the beams.

8. J. A. Wheeler, *Quantum Theory and Measurement* (Princeton, NJ: Princeton University Press, 1984) , p. 184.

9. Ibid, p. 185.

10. Ibid.

11. P. Ball, "Hawking rewrites history . . . backwards," *Nature*, www.nature.com/news/2006/060619/full/news060619-6.

12. S. W. Hawking, "Cosmology from the top-down," in B. Carr, ed., *Universe or Multiverse?* (Cambridge: Cambridge University Press, 2007), p. 91.

13. Stephen Hawking, *A Brief History of Time* (New York: Bantam Books, 1988), p. 139.

14. Ibid., p. 179.

15. Mike Martin, "Hawking: God may play dice after all," *ScienceNet-Daily*, http://www.worldnetdaily.com/news/article.asp?article_id = 27721.

16. J. Ambjørn, J. Jurkiewicz, and R. Loll, "The universe from scratch," *Contemporary Physics* 27 (2006): 103–17; "The self-organizing quantum universe," *Scientific American*, July 2008 (2008): 42–79; "Quantum gravity or the art of building spacetime," in D. Oriti, ed., *Approaches to Quantum Gravity* (Cambridge: Cambridge University Press, 2009), p. 241.

17. A de Sitter geometry is a solution to Einstein's equations for a uni-

verse with positive curvature and devoid of matter. I described de Sitter and anti–de Sitter geometries in chapter 9.

11. ORDER WITHOUT DESIGN

1. These two Hungarian-American scientists were close friends and collaborators. Ulam made significant contributions to the development of the hydrogen bomb, prompting Hans Bethe to say that "after the H-bomb was made, reporters started to call Teller the father of the H-bomb. For the sake of history, I think it is more precise to say that Ulam is the father, because he provided the seed, and Teller the mother, because he remained with the child. As for me, I guess I am the midwife." [Quoted in S. Schweber, *In the Shadow of the Bomb: Bethe, Oppenheimer, and the Moral Responsibility of the Scientist* (Princeton, NJ: Princeton University Press, 2000), p. 166]. Von Neuman contributed fundamental work to quantum mechanics, computer science, and game theory. His brilliance was legendary. The physicist Edward Teller remembers someone stating in jest that (a) Johnny von Neuman can prove anything and (b) anything Johnny proves is correct. After the end of the war, he became, at twenty-nine, the youngest faculty member of the Institute for Advanced Study in Princeton, where they said that von Neuman was without a doubt a demigod but that, through hard study, he'd been able to imitate human beings perfectly.

2. A. Borde and A. Vilenkin, "Violation of the Weak Energy Condition in Inflating Spacetimes," *Physical Review* D56 (1997): 717–23.

3. A. Aguirre and S. Gratton, "Steady-State Eternal Inflation," *Physical Review* D65 (2002): 83507.

4. Or *future*-eternal inflation model, as Alan Guth of MIT calls it. He thinks that the technical assumption that Borde and Vilenkin made in 1997 correcting an earlier paper of theirs that showed that inflationary models must have a singularity in the past does not change those initial conclusions. See A. Guth, "Time since the Beginning," *Astrophysical Ages and Time Scales*, ASP Conference Series 245 (2001).

5. D. Podolsky and K. Enqvist, "Eternal Inflation and Localization on the Landscape," arXiv:hep/0144v3 (2007); A. Aguirre, T. Banks, and M. Johnson, "Regulating Eternal Inflation II: The Great Divide," *Journal of High Energy Physics* (2006): 65.

6. Lee Smolin, "The Status of Cosmological Natural Selection," arXiv: hep/0612185v1 (2006). Smolin asserts in this paper that "both mechanisms

of universe generation [the cosmological natural selection model and eternal inflation] may in fact function in our universe."

7. Larry Abbott, "The Mystery of the Cosmological Constant," *Scientific American*, May 1988, p. 82.

8. P. Creminelli and L. Senatore, "A Smooth Bouncing Cosmology with Scale Invariant Spectrum," arXiv:hep-th/0702165v1 (2007).

9. E. I. Buchbider, J. Khoury, and B. A. Ovrut, "New Ekpyrotic Cosmology," arXiv:hep-th/0702154v4 (2007).

10. His quote reads: "If the universe is really completely self-contained, having no boundary or edge, it would have neither beginning nor end: it would simply be. What place, then, for a creator?" Stephen Hawking, *A Brief History of Time* (New York: Bantam Books, 1988), p. 141.

11. As mentioned in a note to chapter 6, there are intriguing experimental results that are tentatively being interpreted as requiring the existence of additional extra spatial dimensions. In a series of experiments performed at Fermilab in 2007 (the MiniBooNE study), a large number of unexpected muon neutrinos were observed to be transforming into electron neutrinos. While neutrino oscillations have been observed for a long time, the unusual number of these transformations has no conventional explanation. In 2005, Heinrich Päs of the University of Dortmund in Germany, Sandip Pakvasa of the University of Hawaii, and Thomas J. Weiler of Vanderbilt University predicted that a new kind of neutrino may be oscillating in and out of a higher spatial dimension. In light of the new experimental results, several groups of physicists are now working possible mechanisms for the detection of the extra dimensions. Fermilab is planning a new experiment in an attempt to investigate this possibility. The experimental discovery of additional spatial dimensions would provide support for M-theory, which requires the existence of eleven dimensions, ten of space and one of time. See Michelle Maltoni and Thomas Schwetz, "Sterile Neutrino Oscillations after First MiniBooNE Results," *Physical Review* D76 (2007): 093005.

Powers of Ten

Physics deals with quantities that range from the very small to the immensely large. The distance to the nearest galaxy in kilometers, for example, would require many zeros if we were to write it in the conventional way. The mass of an electron in kilograms, a very small number, would require writing 26 zeros after the decimal point. We can avoid these difficulties if we use powers of ten to write these very large and very small numbers.

The product

$$10 \times 10 \times 10 \times 10 \times 10 \times 10$$

where the factor 10 occurs 6 times, can be written as

$$10^6$$

Suppose that we now have the product

$$10 \times 10 \times 10 \times 10 \times 10 \times 10 \times 10 \times 10 \times 10 = 10^9$$

where the factor 10 occurs 9 times. We can write this last product as

$$(10 \times 10 \times 10 \times 10 \times 10 \times 10) \times (10 \times 10 \times 10)$$

or

$$10^6 \times 10^3$$

We can see from the above example that

$$10^6 \times 10^3 = 10^{(6+3)} = 10^9$$

To multiply powers of ten, then, we *add* the exponents.

Suppose that now we want to obtain the result of

$$\frac{10^6}{10^3}$$

which we can write as

$$\frac{10 \times 10 \times 10 \times 10 \times 10 \times 10}{10 \times 10 \times 10}$$

We can then see that

$$\frac{10^6}{10^3} = 10^3$$

To divide powers of ten, we *subtract* the exponents of the numerator and the denominator.

THE FUNDAMENTAL CONSTANTS OF NATURE

The fundamental constants of nature are quantities that appear naturally in the formulation of the theories that describe the universe. These constants are an essential part of the laws of physics.

Constant	*Symbol*	*Value*
Speed of light in vacuum	c	2.998×10^{8} m/s
Charge of electron	e	1.602×10^{-19} C
Mass of electron	m_e	9.109×10^{-31} kg
Gravitational constant	G	6.673×10^{-11} N.m^2/kg^2
Fine structure constant	α	1/137
Planck constant	h	6.626×10^{-34} J.s
Planck length	l_P	1.616×10^{-35} m
Planck mass	m_P	2.176×10^{-8} kg
Planck temperature	T_P	1.417×10^{-32} K
Planck time	t_P	5.391×10^{-44} s
Electric constant	ε_o	8.854×10^{-12} F/m
Magnetic constant	μ_o	$4\pi \times 10^{-7}$ N/A^2
Characteristic impedance at vacuum	Z_o	376.73 Ω

Big and Small Numbers

Hubble constant	$2.3 \times 10^{-18}\ s^{-1}$
Time since the big bang (= 1/Hubble constant)	4.3×10^{17} s
Radius of the observable universe	1.3×10^{26} m
Distance to the Andromeda galaxy	2.8×10^{22} m = 3×10^{6} light-years
Radius of the Milky Way galaxy	5.7×10^{20} m = 6×10^{4} light-years
Radius of the sun	6.96×10^{8} m
Radius of Earth	6.38×10^{6} m
Radius of the moon	1.74×10^{6} m
Distance Earth-sun	1.49×10^{11} m
Distance Earth-moon	3.84×10^{8} m
Number of particles in the universe	10^{80}
Number of particles that make up Earth	10^{50}
Number of particles in the Milky Way galaxy	10^{68}
Mass of the observable universe	2.4×10^{52} kg
Mass of the sun	1.99×10^{30} kg
Mass of Earth	5.98×10^{24} kg
Value of the cosmological constant	$3.9 \times 10^{-36}\ s^{-2}$

GLOSSARY

anthropic principle: The idea that the fundamental constants of nature are consistent with the existence of life in the universe.

anti-de Sitter space: Spacetime with a negative curvature.

antimatter: A type of matter composed of elementary particles with the same physical properties of matter particles except that the electric charge and a property called the *magnetic moment* have opposite sign.

blackbody: An ideal object that absorbs all radiation. It is also a perfect emitter of radiation.

blackbody spectrum: The spectrum emitted by a blackbody. The blackbody spectrum is independent of the material of the body.

black dwarf: A burned-out white dwarf star. The sun will end its life as a black dwarf.

black hole: An object with such strong gravity that not even light can escape.

boson: Force carriers. Elementary particles that carry one of the fundamental forces of nature. Bosons have integer spins.

brane: A membranelike object in M-theory with spatial dimensions ranging from one to ten.

Calabi-Yau manifold: The method used in string theory to compactify the unseen spatial dimensions.

coherent: Light beams vibrating in step.

conservation of energy: The law that states that the total amount of energy you start a process with has to equal what you end up with.

conservation law: The physics law that specifies that a certain quantity cannot be destroyed.

cosmic microwave background (CMB) radiation: The echo of the big bang. The primordial electromagnetic radiation that originated with the quantum fluctuations during the first instants of the universe.

cosmological constant: The only possible term that can be added to Einstein's field equations of general relativity without violating the symmetry of the relativity postulate. The cosmological constant is also known as *vacuum energy*, or *dark energy*.

dark energy: The energy of the vacuum, the cosmological constant.

dark matter: An unknown, invisible substance that can be detected only by its gravitational effects.

de Sitter space: Spacetime with positive curvature.

diffraction: The bending or spreading of a light beam as it passes through a narrow slit or hole.

Doppler effect: The change in the wavelength perceived by a listener who is in motion relative to a source of sound. It also applies to light and other electromagnetic waves.

electron: A fundamental particle that cannot be split and carries negative electric charge. As far as we know, electrons have no size.

energy: The capacity to perform work or the result of doing work.

entropy: A measure of the degree of disorder in a system.

Euclidean time: A mathematical tool in which the time coordinate is expressed in imaginary numbers. Imaginary time behaves like one of the spatial dimensions.

event horizon: The surface of a sphere with a radius equal to the *Schwarzschild radius*.

field: The distortion of space due to the presence of a body that exerts a force on other bodies.

fermions: Elementary particles with half-integer spins. There are two main types of fermions: quarks and leptons.

fractal: A class of complex geometric shapes that exhibits self-similarity at all scales and shows an endless repeated geometry. A Mandelbrot fractal is a well-known example.

fundamental charge: The most fundamental unit of charge is the charge of one electron or one proton. The electric charge on a charged object always occurs in multiples of the fundamental charge.

galaxy: A large island of billions of stars, held together by gravity.

Galilean principle of relativity: The laws of mechanics are the same in all observers in uniform motion. Thus, there is no absolute standard of rest; uniform motion has to have a reference point.

gamma rays: Electromagnetic radiation emitted by certain radioactive nuclei.

gauge symmetry: Physical theories that remain invariant under changes taking place everywhere in the universe are said to obey a global gauge symmetry. When the changes are different at every point in space, the theory is said to obey a local gauge symmetry.

To maintain a local symmetry, in which different changes take place at different points or to different objects, a *compensating* change must take place.

gluon: The carrier of the force between quarks.

graviton: The carrier of the gravitational field.

hadrons: Particles that participate in the strong interaction. Hadrons are not fundamental particles; they have a definite extension. Hadrons that decay into a proton and another stable particle are *baryons*. The remaining hadrons are *mesons*.

Heisenberg uncertainty principle: You cannot determine with complete accuracy the position of a particle and at the same time measure where and how fast it is moving.

Higgs field: The field that participates in the Higgs mechanism.

Higgs mechanism: The spontaneous breaking of the electroweak symmetry that gives mass to the Standard Model particles.

Higgs particle: The particle that carries the Higgs field.

Hubble constant: The constant of proportionality in Hubble law. It gives the rate at which the universe expands. Modern measurements place the value of the Hubble constant at 15 to 30 kilometers per second per million light-years.

Hubble law: The relationship between the recession velocities of distant galaxies and their distances.

interference: The wave that results from the overlap of two wave motions that run into each other. The overlap can be a reinforcement of the waves or a cancellation.

lepton: Particle that interacts via the weak force. All leptons are truly elementary particles, without internal structure. Leptons are classified in three generations, each containing a charged lepton and a neutrino.

microwave: Electromagnetic radiation with wavelengths ranging from about 1 millimeter to 1 meter and, equivalently, with frequencies between 1 gigahertz and 1 terahertz.

M-theory: An extension of superstring theory. M-theory is formulated in an eleven-dimensional spacetime. The full set of equations that describe M-theory has not been discovered yet.

neutrino: A subatomic particle that has no mass or a very tiny amount of mass—we don't yet know which. It plays a role in radioactivity.

neutron: One of the subatomic particles that make up the nucleus of an atom.

neutron star: An object only tens of kilometers across but with a mass larger than that of the sun. Neutron stars are composed mainly of neutrons.

nucleon: Collective name for protons and neutrons.

nucleosynthesis: The process by which all of the elements are generated in thermonuclear reactions that start with simple types of nuclei. The lightest elements were produced during the ten minutes after the big bang. Heavier elements are produced in the interiors of stars.

photon: A quantum of light. The carrier of the electromagnetic field.

Planck's constant: One of the fundamental constants of nature. Planck's constant defines the step size of quanta. Its accepted value is $6.6260693 \times 10^{-34}$ joule second.

Planck length: The natural or Planck unit of length given in terms of Planck's constant, the gravitational constant, and the speed of light. It is equal to 1.6×10^{-35} meter.

Planck mass: The natural or Planck unit of mass, given in terms of Planck's constant, the gravitational constant, and the speed of light. About 10^{-5} grams.

Planck time: The natural or Planck unit of time, given by Planck's constant, the gravitational constant, and the speed of light. Planck time is the time it takes a photon to travel the distance equal to the Planck length. It's about 10^{-44} second.

principle of equivalence: Acceleration and the effects of gravity cannot be distinguished; they are the same phenomenon.

proton: One of the subatomic particles that make up the nucleus of an atom. The proton has positive electric charge.

quanta: Bundles or packets of energy that cannot be split. Light and all electromagnetic radiation are made up of quanta.

quantum chromodynamics (QCD): The part of the Standard Model that explains the behavior of quarks and gluons.

quark: Quarks are believed to be—like leptons—truly fundamental particles. Hadrons are thought to be composed of quarks. There are six flavors or varieties of quarks: *Up* and *down*, *strange* and *charm*, and *bottom* and *top*. Like leptons, quarks are also grouped in generations. The strong force arises from the interaction between quarks. Quarks possess a kind of charge called color charge: red, green, and blue. All hadrons are color neutral or white.

radioactivity: The spontaneous emission of particles or radiation from certain atoms.

red giant: A large, bright star with a low surface temperature. The sun will spend its late years as a red giant, eventually ending as a burned-out star called a black dwarf.

scale-invariant spectrum: A spectrum that can be decomposed into a sum of sinusoidal waves with the same heights.

singularity: Places where the laws of physics break down or do not apply.

spacetime: The combination of the three dimensions of space and one of time. It is needed in relativity because time and space are linked together.

spectrum: The array of the components of the emission of electromagnetic radiation arranged according to wavelength.

spin: A quantum property of elementary particles. The mathematics that describes the spin property resembles that of a spinning ball, except that it only takes certain quantized values.

Standard Model: A description of the behavior of the nongravitational forces and particles that make up the universe.

supergravity: Theory that attempts to unify gravity with the other three forces, with the use of a powerful new gauge symmetry called *supersymmetry* that unites quarks and leptons with messenger particles. Supergravity theories are formulated in more than four spacetime dimensions.

supernova: The last stage in the life of a massive star, with a mass larger than eight suns. A supernova is an exploding star that increases its luminosity by billions of times in a short period.

superstring theory: A theory of the structure of matter that says all elementary particles are represented by tiny vibrating strings. The theory promises to unify all the forces of nature.

supersymmetry: A gauge symmetry that unites quarks and leptons with messenger particles. In supersummetry, the laws of physics are symmetric with the exchange of fermions and bosons.

symmetry: If something remains unchanged after some operation is performed on it, it is said to be symmetric under that operation.

thermodynamics: The study of heat and thermal effects.

tilt: The spectrum in which the heights of the component waves increase or decrease as the frequency changes.

uniform motion: Motion at a constant speed along a straight line.

vacuum energy: The energy stored in the virtual particles that exist in empty space. Also called *dark energy* or the *cosmological constant*.

virtual particle: A particle that exists only for the brief moment, allowed by the Heisenberg uncertainty principle.

wave: A mechanism for the transmission of energy.

wavelength: The distance between two crests or two valleys in a wave.

white dwarf: A small, dense star. One of the final stages in the life of the sun.

FURTHER READING

Barrow, John D. *The Constants of Nature*. London: Jonathan Cape, 2002.

———. *New Theories of Everything*. Oxford: Oxford University Press, 2007.

Barrow, John D., and Frank J. Tipler. *The Anthropic Cosmological Principle*. Oxford: Oxford University Press, 1988.

Calle, Carlos I. *Coffee with Einstein*. London: Duncan Baird, 2008.

———. *Superstrings and Other Things: A Guide to Physics*. London: Taylor and Francis, 2001.

Christianson, Gale E. *In the Presence of the Creator: Isaac Newton and His Times*. New York: Free Press, 1984.

DeWitt, Bryce S., and N. Graham, eds. *The Many-Worlds Interpretation of Quantum Mechanics*. Princeton, NJ: Princeton University Press, 1973.

Glashow, Sheldon L. *The Charm of Physics*. New York: American Institute of Physics, 1991.

Glendenning, Norman. *After the Beginning: A Cosmic Journey through Space and Time*. Singapore: World Scientific, 2004.

———. *Our Place in the Universe*. Singapore: World Scientific, 2007.

Greene, Brian. *The Elegant Universe*. New York: W. W. Norton, 2003.

———. *The Fabric of the Cosmos*. New York: W. W. Norton, 2005.

Guth, Alan. *The Inflationary Universe: The Quest for a New Theory of Cosmic Origins*. New York: Cambridge University Press, 2000.

Einstein, Albert. *The Principle of Relativity*. New York: Dover, 1952.

———. *Relativity: The Special and General Theory*. New York: Crown Publishers, 1961.

Hawking, Stephen. *The Universe in a Nutshell*. New York: Bantam, 2001.

———. *Black Holes and Baby Universes and Other Essays*. New York: Bantam, 1993.

———. *A Brief History of Time*. New York: Bantam, 1988.

———. *The Future of Theoretical Physics and Cosmology*. Cambridge: Cambridge University Press, 2003.

Hawking, Stephen, and Roger Penrose. *The Nature of Space and Time*. Princeton, NJ: Princeton University Press, 1996.

Kaku, Michio. *Parallel Worlds*. New York: Doubleday, 2005.

Kane, Gordon. *Supersymmetry: Unveiling the Ultimate Laws of Nature*. Cambridge, MA: Perseus Publishing, 2000.

Krauss, Lawrence. *Hiding in the Mirror: The Mysterious Allure of Extra Dimensions, from Plato to String Theory and Beyond*. New York: Viking, 2005.

Lederman, Leon, and Christopher Hill. *Symmetry and the Beautiful Universe*. Amherst, NY: Prometheus Books, 2004.

Lightman, Alan, and R. Brower. *Origins: The Lives and Words of Modern Cosmologists*. Cambridge, MA: Harvard University Press, 1990.

Lloyd, Seth. *Programming the Universe: A Quantum Computer Scientist Takes on the Cosmos*. New York: Oxford University Press, 2005.

Pais, Abraham. *Subtle Is the Lord: The Science and Life of Albert Einstein*. Oxford: Oxford University Press, 1982.

Penrose, Roger. *The Road to Reality: The Complete Guide to the Laws of the Universe*. New York: Knopf, 2005.

Randall, Lisa. *Warped Passages: Unravelling the Mysteries of the Universe's Hidden Dimensions*. New York: Ecco, 2005.

Rees, Martin. *Before the Beginning: Our Universe and Others*. Reading, MA: Addison-Wesley, 1997.

Sagan, Carl. *The Varieties of Scientific Experience*. New York: Penguin Press, 2006.

Singh, Simon. *Big Bang: The Origin of the Universe*. New York: HarperCollins, 2004.

Smolin, Lee. *Three Roads to Quantum Gravity*. New York: Basic Books, 2001.

Steinhardt, Paul, and Neil Turok. *Endless Universe: Beyond the Big Bang*. New York: Doubleday, 1997.

Susskind, Leonard. *The Black Hole War*. New York: Little, Brown, 2008.

———. *The Cosmic Landscape: String Theory and the Illusion of Intelligent Design*. New York: Little, Brown, 2006.

Vilenkin, Alex. *Many Worlds in One: The Search for Other Universes*. New York: Hill and Wong, 2006.

Weinberg, Steven. *The Discovery of Subatomic Particles*. New York: Cambridge University Press, 2003.

———. *Dreams of a Final Theory*. New York: Pantheon, 1992.

Wolt, Peter. *Not Even Wrong: The Failure of String Theory and the Search for Unity in the Physical World*. New York: Basic Books, 2006.

Index